Springer Protocols Handbooks

For further volumes:
http://www.springer.com/series/8623

Springer Protocols Handbooks collects a diverse range of step-by-step laboratory methods and protocols from across the life and biomedical sciences. Each protocol is provided in the Springer Protocol format: readily-reproducible in a step-by-step fashion. Each protocol opens with an introductory overview, a list of the materials and reagents needed to complete the experiment, and is followed by a detailed procedure supported by a helpful notes section offering tips and tricks of the trade as well as troubleshooting advice. With a focus on large comprehensive protocol collections and an international authorship, Springer Protocols Handbooks are a valuable addition to the laboratory.

Best Practice in Microbiome Research

Edited by

Simon R. Carding

Food, Microbiome and Health Institute Strategic Programme, Quadram Institute Bioscience, Norwich, UK

Editor
Simon R. Carding
Food, Microbiome and Health Institute
Strategic Programme
Quadram Institute Bioscience
Norwich, UK

ISSN 1949-2448 ISSN 1949-2456 (electronic)
Springer Protocols Handbooks
ISBN 978-1-0716-5008-0 ISBN 978-1-0716-5009-7 (eBook)
https://doi.org/10.1007/978-1-0716-5009-7

This work was supported by UKRI-BBSRC.

This Humana imprint is published by the registered company Springer Science+Business Media, LLC, part of Springer Nature.
The registered company address is: 1 New York Plaza, New York, NY 10004, U.S.A.

Preface

The human microbiome and in particular the gut microbiome are at the leading edge of scientific and biomedical research. The gut microbiome is now regarded as the most significant microbial community in maintaining individual health. Despite the publication of more than 60,000 research articles in the 5-year period 2020 to 2025 (Web of Science) implicating the gut microbiome in both health and disease, there remains a lack of reproducibility among most human microbiome studies.

Identifying universal patterns of gut microbiomes is made difficult by their inherent diversity across individuals, age, gender, and geography together with lifestyle and behavioral factors. This complexity is heightened further by the dynamic nature of the human gut microbiome and by changes in its structure and function that occur over the human lifespan, and by its ability to quickly respond, adapt, and change in response to external factors such as diet, infection, and (oral) medications. The ability to reliably detect and define the microbiome is therefore largely dependent upon the design of human microbiome studies and the parameters and methods of sample collection, processing, and analysis. Bias can be introduced by the lack of standardized protocols and operating procedures (SOPs) for sample collection be they from the same or different sampling sites, by timing and means of transportation, and methods of preservation and storage. In addition, the varied methods for DNA and RNA extraction can introduce additional variability in the content and quality of extracted nucleic acids and the level of contaminating host DNA and RNA and from reagents and kits ("kitomes") themselves that can obfuscate downstream analyses.

A major advancement in human microbiome research is metagenomics-based sequencing which alone or in combination with other "omics technologies" (e.g., transcriptomics, proteomics, and metabolomics) has accelerated and enhanced the ability to interrogate and characterize microbiomes to high levels of detail. However, the use of different sequencing protocols and platforms and the requirement for sophisticated bioinformatics and statistical analysis, which can be computationally intensive and prone to errors can make cross study comparisons challenging or even meaningless.

Equally important are ethical, legal, and social considerations and providing effective safeguards and measures for data ownership, privacy, and security, in addition to considering sources of potential bias such as the choice of study cohorts and criteria for including or excluding study participants that can skew results.

Developing and adopting standardized methods is therefore important and required in all studies attempting to define microbiome signatures and profiles of health and disease states and in attempting to understand its role, if any, in disease causality. This in turn is a major obstacle to translating microbiome research to the clinic. Establishing SOPs for all aspects of human microbiome research from study conception and design through to data analysis and interpretation, sample collection and analysis is a prerequisite for the establishment of biospecimen repositories and large databases, and for validating and comparing results from different studies.

To help promote the adoption of standardized methods for human microbiome research the Quadram Institute is providing a set of protocols that have been developed, optimized, and applied to microbiome studies across the Institute. They cover ethical considerations and study design in addition to recommendations for standardizing sample collection, transport, storage, and processing. The inclusion of protocols for sampling and analysis of microbiomes from sites other than the gut and for analyzing the virome and mycobiome are unique features and emphasizes the importance and need for inclusive and integrative microbiome studies that go beyond the analysis of a single constituent and the prokaryome. These protocols highlight critical concepts and key features, and in doing so serve as a valuable benchmark. It is important to note that some protocols, particularly those involving automation and informatics, are constantly evolving, and being improved.

These protocols are being made freely available to all researchers to enable and encourage their adoption and use and for the wider benefit of human microbiome research.

Norwich, UK ***Simon R. Carding***

Contents

Contributors

EVELIEN M. ADRIAENSSENS • *Food, Microbiome and Health Institute Strategic Programme, Quadram Institute Bioscience, Norwich, UK; Microbes and Food Safety Institute Strategic Programme, Quadram Institute Bioscience, Norwich, UK*

JENNIFER AHN-JARVIS • *Food, Microbiome and Health Institute Strategic Programme, Quadram Institute Bioscience, Norwich, United Kingdom*

DAVE BAKER • *Food, Microbiome and Health Institute Strategic Programme, Quadram Institute Bioscience, Norwich, UK*

GEMMA BEASY • *Food, Microbiome and Health Institute Strategic Programme, Quadram Institute Bioscience, Norwich, UK*

SIMON R. CARDING • *Food, Microbiome and Health Institute Strategic Programme, Quadram Institute Bioscience, Norwich, UK*

RYAN COOK • *Food, Microbiome and Health Institute Strategic Programme, Quadram Institute Bioscience, Norwich, UK*

MARIANNE DEFERNEZ • *Norwich Medical School, University of East Anglia, Norwich, United Kingdom*

ANTHONY DUNCAN • *Food, Microbiome and Health Institute Strategic Programme, Quadram Institute Bioscience, Norwich, UK*

RIK HAAGMANS • *Food, Microbiome and Health Institute Strategic Programme, Quadram Institute Bioscience, Norwich, UK*

ANTONIETTA HAYHOE • *Food, Microbiome and Health Institute Strategic Programme, Quadram Institute Bioscience, Norwich, UK*

SAMUEL J. HAYNES • *Food, Microbiome and Health Institute Strategic Programme, Quadram Institute Bioscience, Norwich, UK*

FALK HILDEBRAND • *Food, Microbiome and Health Institute Strategic Programme, Quadram Institute Bioscience, Norwich, UK*

SARAH HUGHES • *Food, Microbiome and Health Institute Strategic Programme, Quadram Institute Bioscience, Norwich, UK*

STEVE JAMES • *Food, Microbiome and Health Institute Strategic Programme, Quadram Institute Bioscience, Norwich, UK*

WING KOON • *Food, Microbiome and Health Institute Strategic Programme, Quadram Institute Bioscience, Norwich, UK*

REVATHY KRISHNAMURTHI • *Food, Microbiome and Health Institute Strategic Programme, Quadram Institute Bioscience, Norwich, UK*

VIET THANH LE • *Food, Microbiome and Health Institute Strategic Programme, Quadram Institute Bioscience, Norwich, UK*

GWÉNAËLLE LE GALL • *Norwich Medical School, University of East Anglia, Norwich, United Kingdom*

CARA-JANE MOSS • *Food, Microbiome and Health Institute Strategic Programme, Quadram Institute Bioscience, Norwich, UK*

EZGI ÖZKURT • *Food, Microbiome and Health Institute Strategic Programme, Quadram Institute Bioscience, Norwich, UK*

MICHAEL D. PAXHIA • *Food, Microbiome and Health Institute Strategic Programme, Quadram Institute Bioscience, Norwich, United Kingdom*

SARAH PHILLIPS • *Food, Microbiome and Health Institute Strategic Programme, Quadram Institute Bioscience, Norwich, UK*
ALISE J. PONSERO • *Food, Microbiome and Health Institute Strategic Programme, Quadram Institute Bioscience, Norwich, UK*
GEORGE M. SAVVA • *Core Science Resources, Quadram Institute Bioscience, Norwich, UK; Core Science Resources, Quadram Institute Bioscience, Norwich, United Kingdom*
ILIANA SERGHIOU • *Food, Microbiome and Health Institute Strategic Programme, Quadram Institute Bioscience, Norwich, UK*
KATARZYNA SIDORCZUK • *Food, Microbiome and Health Institute Strategic Programme, Quadram Institute Bioscience, Norwich, UK*
JUDIT TALAS • *Food, Microbiome and Health Institute Strategic Programme, Quadram Institute Bioscience, Norwich, UK*
ANDREA TELATIN • *Food, Microbiome and Health Institute Strategic Programme, Quadram Institute Bioscience, Norwich, UK*
SUMEET K. TIWARI • *Food, Microbiome and Health Institute Strategic Programme, Quadram Institute Bioscience, Norwich, UK*
SILAS TRILLER • *Food, Microbiome and Health Institute Strategic Programme, Quadram Institute Bioscience, Norwich, UK*
RACHEL WATT • *Food, Microbiome and Health Institute Strategic Programme, Quadram Institute Bioscience, Norwich, UK*

Chapter 1

Research Ethics

Antonietta Hayhoe and Jennifer Ahn-Jarvis

Abstract

Although research ethics governance varies by geographical location, the core principles of the ethical review process remain relatively consistent and play a critical role in mitigating risks associated with microbiome research. This chapter reviews common practices, including adherence to the International Council for Harmonisation's Guideline for Good Clinical Practice (ICH GCP), independent review, oversight structures, participant protection, informed consent, and public involvement. It also addresses the unique ethical responsibilities of microbiome researchers, such as promoting health equity and ensuring data protection.

Key words Good Clinical Practice, Research ethics committee, Regulatory requirements, National laws, Informed consent, Tissue banking

1 Introduction

The quality of microbiome research heavily depends on conducting studies involving research participants, human tissue, and data in accordance with fundamental ethical principles and relevant regulations. This includes clearly defining the purpose of the research, adhering to ethical standards such as protecting participants' rights and well-being and handling donated biological materials and data according to established procedures that ensure safety and privacy during collection, storage, and use. Compliance with national and international laws and guidelines is also essential. Adherence to these standards directly influences the quality and reliability of research outcomes [1, 2]. Country-specific regulatory requirements for clinical research can be accessed through ClinRegs (https://clinregs.niaid.nih.gov/), an online database hosted by the National Institute of Allergy and Infectious Diseases [3].

Simon R. Carding (ed.), *Best Practice in Microbiome Research*, Springer Protocols Handbooks,
https://doi.org/10.1007/978-1-0716-5009-7_1, © The Author(s) 2026

2 Ethical Framework

2.1 Governance

Specifically in the United States (US), the ethics review process is governed by the Common Rule and overseen by Institutional Review Boards (IRBs). The Common Rule was adopted in 1991 during the revision of Title 45 of the Code of Federal Regulations, Part 46 (Public Welfare), Subparts A, B, C, and D, which pertain to behavioral research involving participants in the US [4]. It underwent its first significant revision in 2018 to enhance participant protections and reduce administrative burdens [5]. In Europe, national laws, and a combination of European Union (EU)-wide regulations, such as the General Data Protection Regulation (GDPR) for privacy and the Clinical Trials Regulation for clinical research, set overarching standards. However, each EU member state retains significant autonomy in implementing these regulations, resulting in variations in governance practices [6]. In the United Kingdom (UK), the UK Policy Framework for Health and Social Care Research developed by the Health Research Authority (HRA), in collaboration with health departments across the UK, provides a unified set of principles for all health and social care research [7].

2.2 Oversight

Despite recent changes to the Common Rule, the US ethics review system, known as the Institutional Review Board (IRB), continues to be established independently within institutions. There are approximately 2300 IRBs, which are overseen by the Food and Drug Administration (FDA) and the Office for Human Research Protections (OHRP) [8]. This contrasts with European countries, which have moved toward a centralized system.

In England, Research Ethics Committees (RECs) are responsible for reviewing research proposals to ensure the rights, safety, dignity, and well-being of participants are protected, while also supporting ethical research that benefits society [9]. RECs typically include up to 15 members, with one-third being "lay" members, individuals who have no primary professional interest in research or healthcare. In the UK, Public and Patient Involvement (PPI) has become an essential and mandatory component of the REC review process since December 2024. This change, led by the Health Research Authority (HRA) and applied across the UK, ensures that RECs can better address the concerns and needs of participants, thereby enhancing the relevance and acceptability of research studies [10].

3 Informed Consent

Informed consent in human microbiome research presents unique ethical, legal, and social challenges and must be tailored with specific considerations. Due to the evolving nature of microbiome research, uncertainties regarding the risks and benefits of the microbiome will persist at the time of consent. Participants may need to accept a degree of uncertainty as part of their informed decision-making [11–13]. Furthermore, new discoveries may necessitate changes to consent processes to protect both prospective and already enrolled participants. Participants should be clearly informed about how their samples will be collected, used, and stored and whether their samples will be anonymized or linked to their identities. Microbiome research often involves detailed consent forms that address the storage of biological specimens, privacy risks, and physical risks associated with invasive sampling methods (e.g., collection from multiple body sites). Privacy risks must be explicitly communicated in the consent forms, especially since microbiome data can potentially reveal sensitive health information [14].

3.1 Comparing Broad and Study-Specific Consent: Key Differences and Applications

Broad consent enables researchers to use microbiome data for unforeseen studies, which is particularly important in this field of research. However, whether participants can truly provide general consent for unspecified future research is controversial. Broad consent offers researchers the flexibility to collect and store identifiable biospecimens or private information for secondary research without requiring re-consent for each new study [15].

In the US, broad consent was introduced with the 2018 revisions to the Common Rule as an alternative to study-specific consent [16]. It is especially useful for biobanking or when future research directions are uncertain. However, protections for participant autonomy and privacy must be in place, particularly in the context of data protection and usage. Participants must also be informed about the uncertain nature of potential risks [15].

In contrast, specific consent is tailored to a particular study and requires detailed information about its purpose, procedures, risks, and benefits. This type of consent enhances transparency and supports participant autonomy by providing explicit details about the nature and objectives of each study, which is important for fostering trust [17]. Specific consent aligns with the ethical principles of informed consent. However, it applies only to the study or set of studies for which it was initially granted. If new research opportunities arise or secondary use of biological samples is needed, researchers must seek additional consent from participants [18], which can be a limitation for long-term research programs.

3.2 Best Practices for Informed Consent

As our understanding of the microbiome expands, the informed consent process for microbiome research must evolve alongside the science, while continuing to prioritize participant autonomy, privacy, and justice.

1. Adapt Consent Language

   Simplify consent forms and use culturally appropriate language to ensure participants fully understand the information provided [12].

   Provide sufficient time and opportunities for participants to discuss their decisions with trusted individuals or community representatives [12].

   Adapt consent processes to accommodate the needs of vulnerable populations, including Indigenous groups [19, 20].

2. Consent Framework to be Community-Centered

   Engage the public in codesigning research to ensure its relevance and promote mutual benefits [21].

   Empower participants by giving them decision-making authority over how their data is used and shared [21].

3. Avoid Assent

   Avoid passive or group assent (e.g., consent from gatekeepers) to uphold participant autonomy and ensure individuals provide full consent when they are ready.

4. Transparency on Risks and Benefits

   Clearly outline the potential risks (e.g., privacy breaches) and benefits (e.g., health improvements) of participation to ensure informed decision-making [11].

5. Equitable Benefit Sharing

   Ensure health equity and provide tangible benefits for participants and their communities, such as access to health interventions or study findings [21].

4 Protection of Research Participants

While early microbiome research laid the foundation for our current understanding of the field, it has faced considerable criticism for excluding vulnerable groups and failing to address ethical concerns. These exclusions, often justified on scientific grounds, can introduce bias into microbiome research [21]. Vulnerable populations in research include children, elderly individuals, prisoners, people with intellectual or mental impairments, those who are economically or educationally disadvantaged, pregnant women and fetuses, and individuals with language barriers [22]. These groups may face challenges in understanding the complexities of microbiome research due to educational, cultural, linguistic, or cognitive barriers. Indigenous microbiome research presents additional ethical complexities, including issues of data ownership, health equity, and the need for culturally tailored informed consent. Ethical frameworks should be developed to ensure that Indigenous communities retain control over research decisions and data [21].

5 Data Protection

Advances in microbiome methodologies have led to the collection of highly sensitive data that can potentially identify individuals. Human microbiome data, when combined with genomic and medical information, can reveal an individual's disease susceptibility and lifestyle habits [17]. A breach of privacy involving such information could significantly impact participants' rights, including areas such as medical insurance and employment. Therefore, robust data protection measures are essential to safeguard participant identities.

Researchers and ethics review committees must strive to preserve participant diversity while ensuring that ethical concerns related to race- or ethnicity-based conclusions are addressed and mitigated [23]. It is crucial for research organizations to implement standardized procedures for collecting, storing, and using data to ensure privacy and compliance with regulations governing sensitive biological information. These include the General Data Protection Regulation (GDPR) enforced in the EU [24], the Data Protection Act 2018 (DPA) in the UK [25], and key federal laws in the US such as the Health Insurance Portability and Accountability Act (HIPAA), which ensures that patient data is securely stored and transmitted. HIPAA-compliant infrastructure is especially important for microbiome testing services, as these tests often involve sensitive genetic and health-related data [26].

6 Empowering Communities in Microbiome Research

Ethical frameworks should prioritize community involvement in research decisions and equitable benefit sharing. Cultural and religious expectations must be respected during sample collection, and transparency is essential regarding how profits from derived products are allocated. Approaches to PPI and engagement (PPIE) have primarily been developed for clinical research [27]. However, a recent report by the UK Medical Research Council (MRC) highlighted the growing importance of public involvement in non-clinical and biomedical research [28]. In the US, PPIE is increasingly recognized as essential for enhancing the relevance and quality of health research. The All of Us Research Program by the National Institutes of Health (NIH) aims to collect health data from diverse populations to improve health outcomes, with a strong emphasis on participant engagement [29]. Similarly, the Patient-Centered Outcomes Research Institute (PCORI) funds research that involves patients and stakeholders throughout the research process, including through patient advisory boards and community engagement activities [30]. PCORI's Engagement Awards Program further supports projects that integrate patients, caregivers, clinicians, and other stakeholders, fostering a community of engaged contributors [31]. Despite its growing importance, PPIE still faces challenges such as resource allocation, effective communication, and sustaining long-term engagement. A long-standing disconnect has existed between scientists and the communities they aim to serve. Public engagement groups help bridge this gap, enabling researchers to bring their work back to the community. Microbiome researchers should ensure that public voices are central to all stages of research. In 2019, the UK National Institute for Health Research (NIHR) established the UK Standards for Public Involvement in Research to enhance the quality and consistency of public involvement in health and social care research. These standards cover six key areas: Inclusive Opportunities, Working Together, Support and Learning, Communications, Impact, and Governance [32]. However, PPIE remains limited in the early stages of research. A recent scoping review by Mathieson found that engaging various partners, including healthcare professionals and lay members, in implementation research is resource- and time-intensive [33]. Nevertheless, it enhances the cultural appropriateness and sustainability of community-based programs. Improved reporting and evaluation in peer-reviewed literature could help clarify the impact of PPIE [33]. Raising awareness within the scientific community about the importance of public engagement is crucial. Without it, researchers risk overlooking the questions and concerns that matter most to the public. PPIE ensures that research remains relevant, incorporating invaluable insights from those with lived experience. Public contributions can significantly enhance microbiome study design, address ethical concerns, and improve both recruitment and retention rates.

Acknowledgements

This work was funded by the Biotechnology and Biological Sciences Research Council (BBSRC) through the Institute Strategic Programme awards Gut Microbes and Health BB/R012490/1 and Food, Microbiome and Health [BB/X011054/1] and their constituent project(s).

References

1. Hearnshaw H (2004) comparison of requirements of research ethics committees in 11 European countries for a non-invasive interventional study. Br Med J 328(7432): 140–141. https://doi.org/10.1136/bmj.328.7432.140
2. Goodyear-Smith F, Lobb B, Davies G, Nachson I, Seelau SM (2002) International variation in ethics committee requirements: comparisons across five westernised nations. BMC Med Ethics 3:1–8. https://doi.org/10.1186/1472-6939-3-2
3. Kagan J, Goodman GN, Oh R, Whitworth D, Gladden D, Graves G, Nguyen A, Shrestha O, Burge G, Smolskis M, Andrews J, Cramer B, Lane HC (2021) NIAID ClinRegs—a public database of country clinical research regulatory and ethics requirements: design and utilization analysis. Clin Infect Dis 73(7):1296–1301. https://doi.org/10.1093/cid/ciab505
4. 45 CFR 46 (2024) US department of health and human services. Office for Human Research Protections, Washington, D.C.. https://www.hhs.gov/ohrp/regulations-and-policy/regulations/45-cfr-46/index.html
5. Gartel G, Scuderi H, Servay C (2020) Implementation of common rule changes to the informed consent form: a research staff and institutional review board collaboration. Ochsner J 20(1):76–80. https://doi.org/10.31486/toj.19.0080
6. Kromar S, Kassas R (2025) Navigating EU clinical trials: adapting to a new era of regulations. J Cancer Policy 43:100563. https://doi.org/10.1016/j.jcpo.2025.100563
7. Health Research Authority (2017) UK policy framework for health and social care research. Health Research Authority. https://www.hra.nhs.uk/planning-and-improving-research/policies-standards-legislation/uk-policy-framework-health-social-care-research/
8. Institutional Review Boards: Actions Needed to Improve Federal Oversight and Examine Effectiveness (2023) US Government Accountability Office, Washington, D.-C. https://www.gao.gov/products/gao-23-104721
9. Health Research Authority (2020) Governance arrangements for research ethics committees, 2020 edn. Health Research Authority. https://www.hra.nhs.uk/planning-and-improving-research/policies-standards-legislation/governance-arrangement-research-ethics-committees/
10. Health Research Authority (2024) Participant information quality standards. Health Research Authority. https://www.hra.nhs.uk/planning-and-improving-research/research-planning/participant-information-quality-standards/
11. McGuire AL, Colgrove J, Whitney SN, Diaz CM, Bustillos D, Versalovic J (2008) Ethical, legal, and social considerations in conducting the human microbiome project. Genome Res 18(12):1861–1864. https://doi.org/10.1101/gr.081653.108
12. Ma Y, Chen H, Lan C, Ren J (2018) Help, hope and hype: ethical considerations of human microbiome research and applications. Protein Cell 9(5):404–415. https://doi.org/10.1007/s13238-018-0537-4
13. Rhodes R (2016) Ethical issues in microbiome research and medicine. BMC Med 14(1): 156. https://doi.org/10.1186/s12916-016-0702-7
14. Shin A, Xu H (2022) Privacy risks in microbiome research: public perspectives before and during a global pandemic. Ethics Human Res 44:2–13. https://doi.org/10.1002/eahr.500132
15. Steinsbekk KS, Kåre Myskja B, Solberg B (2013) Broad consent versus dynamic consent in biobank research: is passive participation an ethical problem? Eur J Hum Genet 21(9): 897–902. https://doi.org/10.1038/ejhg.2012.282
16. Maloy JW, Bass PF (2020) Understanding broad consent Ochsner journal 20(1): 81–86. https://doi.org/10.31486/toj.19.0088
17. Callaway E (2015) Microbiome privacy risk. Nature 521(7551):136. https://doi.org/10.1038/521136a
18. Kling S (2019) Ethics and the microbiome. Curr Allergy Clin Immunol 32(2): 101–103. https://hdl.handle.net/10520/EJC-168045ed96
19. UK Research and Innovation (2025) Research with potentially vulnerable people. https://www.ukri.org/councils/esrc/guidance-for-applicants/research-ethics-guidance/research-with-potentially-vulnerable-people/
20. National Institutes of Health (2025) Vulnerable and other populations requiring additional protections. U.S. Department of Health and Human Services. https://grants.nih.gov/policy-and-compliance/policy-topics/human-subjects/policies-and-regulations/vulnerable-populations

21. Jones S (2024) An ethical way forward for indigenous microbiome research. Nature. https://doi.org/10.1038/d41586-024-02792-w
22. U.S. Department of Health and Human Services (n.d.) Vulnerable populations. Office for Human Research Protections. https://www.hhs.gov/ohrp/regulations-and-policy/guidance/vulnerable-populations/index.html
23. Hoffmann DE, von Rosenvinge EC, Roghmann MC, Palumbo FB, McDonald D, Ravel J (2024) The DTC microbiome testing industry needs more regulation. Science 383(6688): 1176–1179. https://doi.org/10.1126/science.adk4271
24. European Parliament & Council of the European Union (2016) Regulation (EU) 2016/679 of the European Parliament and of the council of 27 April 2016 on the protection of natural persons regarding the processing of personal data and on the free movement of such data, and repealing directive 95/46/EC (general data protection regulation). Off J European Union L 119:1–88. https://eur-lex.europa.eu/eli/reg/2016/679/oj/eng
25. United Kingdom (2018) Data Protection Act 2018. https://www.legislation.gov.uk/ukpga/2018/12/contents
26. U.S. Department of Health and Human Services (1996) Health insurance portability and accountability act of 1996 (HIPAA), Public Law, pp 104–191. https://www.hhs.gov/hipaa/for-professionals/privacy/laws-regulations/index.html
27. National Institute for Health and Care Research (2025) Practical approaches to patient & public involvement in research. https://www.dementiaresearcher.nihr.ac.uk/practical-approaches-to-patient-public-involvement-in-research/
28. Medical Research Council (2024) MRC NMGN public partnership strategy 2024–2027. National Mouse Genetics Network https://nmgn.mrc.ukri.org/working-groups/public-partnership/mrc-nmgn-public-partnership-strategy
29. National Institutes of Health (n.d.) All of Us Research Program. https://www.nih.gov/taxonomy/term/846/all
30. Patient-Centered Outcomes Research Institute (2025) The PCORI strategic plan: national priorities for health. https://www.pcori.org/about/pcori-strategic-plan/pcori-strategic-plan-national-priorities-health
31. Patient-Centered Outcomes Research Institute (n.d.) Eugene Washington PCORI Engagement Award Program. https://www.pcori.org/engagement-research/eugene-washington-pcori-engagement-award-program
32. National Institute for Health Research. (2019) UK standards for public involvement in research. https://www.nihr.ac.uk/news/nihr-announces-new-standards-public-involvement-research
33. Mathieson A, Brunton L, Wilson PM (2025) The use of patient and public involvement and engagement in the design and conduct of implementation research: a scoping review. Impl Sci Commun 6:42. https://doi.org/10.1186/s43058-025-00725-w

Chapter 2

Study Design Considerations

George M. Savva, Alise J. Ponsero, Evelien M. Adriaenssens, Falk Hildebrand, and Simon R. Carding

Abstract

This chapter provides guidelines and advice for designing and setting up observational and experimental microbiome studies, with a primary focus on human-associated microbiome research. The main planning steps are establishing protocols, defining research questions, study design, and evaluating resources. Attention is given here to technical considerations such as quality control and data management strategies. Additional resources are signposted throughout.

Key words Study types, Reproducibility, Integrity, Covariates, Controls, Inclusion exclusion criteria, Sampling, Power calculation

1 Introduction

This chapter provides guidance for researchers planning microbiome studies. While the focus is on human gut microbiome studies, many of these principles can be applied to microbiome studies in other contexts.

While principles of good study design apply across all biomedical research, microbiome studies present unique challenges. The complex, individual, dynamic nature of microbial communities, their sensitivity to environmental factors, and the technical complexities of their measurement necessitate specific design considerations. This chapter addresses both general principles of clinical study design for those new to human research and specific adaptations required for studying microbiomes.

There are many textbooks, reviews, and guidelines for clinical and epidemiological research [1–3] and for microbiome research in specific contexts [4–7], and researchers are encouraged to consult relevant methodological literature early in the design phase, as well as this general guide.

Simon R. Carding (ed.), *Best Practice in Microbiome Research*, Springer Protocols Handbooks,
https://doi.org/10.1007/978-1-0716-5009-7_2, © The Author(s) 2026

1.1 Study Protocols, Standard Operating Procedures, and Regulations

Studies involving human participants need a protocol, an agreed document that is ethically reviewed, that investigators will follow, and that will define most aspects of study conduct. Obtaining a protocol template from the study sponsor or regulator will help identify the aspects of study design that need to be addressed (see, e.g., [8]).

Depending on the study aims and jurisdiction, there may be additional regulatory guidelines set by, for example, the US Food and Drug Administration (FDA) or European Medicines Agency (EMA). These may also govern study design and implementation. Whether the study falls under a regulated category broadly depends on whether the study is experimental or of a therapeutic product. Treatments targeting the microbiome, for example, fecal microbial transplantation (FMT) or probiotics, can fall into a grey area between therapeutics and nutritional supplements, and how they are classified may vary by jurisdiction, purpose of study, nature of intervention, and the effects being tested. Up-to-date guidance on trial regulation should be sought at the earliest opportunity.

For nonregulated studies, such as observational microbiome research or exploratory academic studies, it is still important to follow best practices. Ensuring Good Clinical Practice (GCP) compliance [9], obtaining ethical approvals, and planning and documentation of objectives, study design, participant selection and recruitment, microbiome sampling, sequencing methods, bioinformatics workflow, and statistical analysis are all important.

Standard operating procedures (SOPs) then set out step-by-step instructions for the processes outlined in the protocol that are required for study delivery. These help to ensure the consistency of methods across a study, quality control, and reproducibility between studies. SOPs should cover recruitment, sample collection, storage and analysis, and data management (see Chaps. 3, 4, 5, 6, 7, 8, 9, 10, 11, 12, and 13 for SOPs and guidance).

1.2 Feasibility and Pilot Studies and Study Resources

If there are uncertainties around the feasibility of any aspect of a study, then pilot and feasibility studies can help to reassure funders and regulatory bodies [10]. Feasibility studies primarily assess whether the proposed research methods are practical and likely to be successful, focusing on participant recruitment strategies, sample collection techniques, methods for the collection of outcome measures, exposures and covariates, storage conditions, sequencing protocols, and data analysis pipelines.

Pilot studies test these methods in small cohorts to refine protocols, assess recruitment rates and participant retention and adherence, and can provide preliminary indications of effect sizes and inter- and intraindividual variability in microbiome composition within study populations to support the design of larger trials.

Resource limitations must be considered throughout study development including personnel, equipment, facilities, and time.

Full cost implications of the chosen methods, including participant recruitment and sample collection, sequencing platform selection, sequencing depth, sample and data storage, bioinformatic and statistical analysis, and publication costs, need to be considered. Study timelines must align with the funding and professional constraints, and the staffing and clinical limitations within the research setting.

Honest discussion with clinical collaborators and those with previous experience of conducting research in the proposed setting is important to understand what is realistic in terms of recruitment, drop out, participants' willingness to undertake measurements, and the clinical staffing required for study delivery.

Even so, issues with research studies may become apparent only when the study starts, and a plan for proactive monitoring of the study progress and possible adaptation of the study protocol, while ensuring the study aims are not compromised, is important.

1.3 Research Study Teams and Planning Checklists

Clinical research studies are complex, requiring the coordination of many different individuals and organizations. Study teams including scientific experts, methods specialists, clinical collaborators, and participant group representatives should be formed early and regularly consulted throughout the design and conduct of the study.

Checklists to help with study planning, design, conduct, and reporting help ensure that key areas are not overlooked, and potential difficulties can be anticipated before a study begins. The UK Research Integrity Office (UKRIO) has produced a nontechnical checklist covering essential elements of good practice covering the entire life cycle of any research study [11].

Reporting checklists, such as those listed below, should be followed when preparing findings for publication and to help identify elements that should be considered at the study design phase:

- The STROBE (Strengthening the Reporting of Observational Studies in Epidemiology) checklist is valuable for observational studies [12].
- The CONSORT (Consolidated Standards of Reporting Trials) applies to randomized controlled trials [13].
- The STORMS (Strengthening The Organizing and Reporting of Microbiome Studies) [14] and the STROBE-metagenomics [15] checklists provide additional specific standards for microbiome research reporting.
- The EQUATOR Network (Enhancing the QUAlity and Transparency Of health Research) more generally provides comprehensive resources for various study designs in different fields [16].

Note the requirements for reporting on patient flows throughout the study and be sure to set up the record keeping in the selection and recruitment process that this will require.

2 Methods

2.1 Setting the Research Question

The aims of research studies, that is, the research questions that they aim to answer, govern their design. Conversely, practical considerations affect the feasibility of the research questions to be addressed [17]. Since decisions around all other aspects of research design rest on the specific questions that the study aims to answer, it is important that these are specified as precisely as possible and agreed upon within the research team.

2.1.1 Types of Research Questions

While a focus is often on designs to answer causal or mechanistic questions, research studies can have other goals, including diagnostics and prediction, description of the natural history of the participant group, or they might have methodological goals. Even in these cases, or if the goals are exploratory or hypothesis-generating, it should be possible to state a research question so that it will be clear when it has been answered hence the study has been successful. Regardless of the type of study, the research questions should be critically assessed:

- Is it specific and measurable?
- Does it address a useful gap in current knowledge?
- Can it be ethically answered with available methods and resources?

Types of research questions include the following:

1. *Descriptive/Exploratory* questions aim to describe populations and find associations or differences between groups at an observational level. For example, what is the structure of the microbiome found within a specific population at specific sites, and how does this evolve over time?
2. *Causal questions* aim to estimate or test causal associations between different factors. For example, to what extent do features of the microbiome either affect or are they affected by another factor? More complex causal questions include questions of mediation or moderation, for example, asking whether the microbiome either acts on the causal path between a treatment and a health outcome or modifies the effect of a treatment.
3. *Predictive* questions aim to use the microbiome to predict a future outcome, e.g., can the microbiome be used as a diagnostic or predictive biomarker?

4. *Methodological, feasibility, or pilot studies* aims to address specific questions related to study design, methodology, and potential challenges before embarking on large-scale investigations to answer other questions.

Simple clinical research questions, particularly causal questions using experimental designs, can be formulated using the PICOT framework [18]. Carefully documenting a question using a framework like this ensures that key elements of the question are well enough specified and agreed upon within a research team. The PICOT framework requires a specific description of:

- **P**opulation (to whom should the results apply).
- **I**ntervention (what is the experimental treatment).
- **C**omparator (what is the experimental treatment being compared to?)
- **O**utcome (what is the primary endpoint that we are trying to manipulate).
- **T**imeline (over what time course?)

2.1.2 Literature Reviews

A thorough literature review before designing a study will identify related completed or ongoing studies, their methods, and justifications, and ensure that the research questions are genuinely unanswered. Literature reviews should be structured and reproducible to ensure relevant literature is included and should avoid cherry-picking references to support the case for any proposed research.

Systematic literature reviews are research studies and aim to answer research questions through combining and comparing existing work. These can be enormously valuable and highly impactful studies and should be carefully conducted according to guidelines from, for example, the Cochrane Collaboration [19]. Authors should consider whether a systematic review and meta-analysis of existing work might be a better use of limited resources than generating further primary evidence to address their question [20].

2.2 Select a Study Design

2.2.1 Types of Studies and Strength of Evidence

Research studies can be broadly classified into two groups: (a) those in which the investigators manipulate experimental conditions to observe the outcome (experimental or intervention studies) and (b) those in which conclusions are drawn through observation alone (non-interventional or observational studies).

While this chapter outlines basic designs and some of the specific considerations related to microbiome research, it is important to be aware of the general literature on study design that is specific to the type of study being planned. Many books cover research design issues for laboratory biologists and those working with animal models [21–23], for human clinical trialists [24], and for epidemiology [2, 25].

2.2.2 Experimental Studies

Experimental studies are used to answer mechanistic questions and test the efficacy or effectiveness of interventions. They provide the most powerful evidence of causal relationships between interventions and outcomes but can be difficult and expensive to conduct, are limited in scope, and can raise considerable ethical questions, particularly for human studies.

The most used experimental designs for human participants are parallel group and crossover randomized controlled trials. Both designs use control groups or control periods and rely on random allocation of treatments. Randomization is important for reducing bias and identifying useful effect estimates [26].

Experimental designs without controls should be avoided, if possible, because following an uncontrolled study, it is very difficult to estimate the effect of a treatment that is independent of the effect of other aspects of study participation, the passage of time, or regression to the mean. Contemporaneous control conditions in which participants are treated identically except for random allocation to the treatments under investigation are the most reliable way of removing biases from these sources.

In parallel group studies, comparisons are made between two or more groups of participants who are allocated to different treatments, while in crossover studies each participant will undergo more than one different treatment, each participant effectively acting as their own control, with the effect estimate based on comparisons within participants.

Other randomized experimental designs might be considered, including designs with wait-list controls or stepped-wedge designs, N-of-1 trials, or cluster randomized designs. These are rarely used for microbiome studies, although cluster randomization is appropriate when individuals must be randomly allocated to treatments in groups (e.g., for a geographical or practice-level intervention) [27]. Factorial designs can be efficient and should be considered if several different interventions are being investigated [28].

Crossover Studies

Cross-over studies enable comparisons of treatments by sequentially administering more than one treatment to each participant, with the order of treatments randomized. By comparing effects within individuals, and so removing between-participant variation from effect estimates, cross-over studies are potentially very efficient, and so can use fewer participants than parallel designs. However, crossover studies rely on treatment effects being reversible on cessation of treatment. The plausibility of this will depend on the intervention and outcome being studied.

For short-term (e.g., dietary) interventions aiming to modify the microbiome composition, a wash-out period of several weeks between cessation of one treatment and beginning another should be used to avoid carry-over effects across treatments [29]. For other interventions, e.g., FMT, it is unlikely that reversing any effect is possible, and so crossover designs cannot be used.

Cross-over studies can greatly increase the total study duration and the burden for each participant compared to parallel designs with consequences for ethics, recruitment, and retention of participants, although the total number of assessments required for the study will still often be reduced compared to a parallel design, making them cost-effective if feasible.

Parallel Group Studies

Parallel group studies, in which participants are each allocated a single treatment, are simpler conceptually but need more participants, since considerable variation in outcomes between participants will affect the estimation of treatment effects.

This can be mitigated to some extent by pre-treatment measurement of microbiomes and other outcomes, with the pre-treatment values then used as statistical controls in analysis. Since measures of microbiomes are often highly correlated within individuals over time, using pre-treatment measures as covariates might substantially improve precision and power. The high chance of personalized responses to microbiome-based interventions means that, as well as estimating average treatment effects, as is typical with many research designs, microbiome studies are most useful if designed to estimate individual responses to treatment. In practice, this means more frequent sample collection, repeat measures within participants within periods wherever possible, and enough participants to enable the estimation of random treatment effects [30]. If subgroup analyses are of interest, then these should be anticipated in the design with appropriate participant selection and sample size.

Treatment durations and the length of follow-up are important considerations. These should be informed by a theoretical understanding of the specific mechanism of the intervention and its effect on the outcome, for example, some dietary impacts on the microbiome can take place almost immediately, while some may not be evident for months [29, 30]. For explanatory studies, it is important to maximize the chance of observing any theorized effect, while in pragmatic studies the study duration should reflect likely intervention periods in a real-world implementation.

2.2.3 Observational Studies

Observational studies (sometimes called natural experiments) do not use any manipulation of participants. They can be simpler and usually cheaper to conduct than clinical trials, although they are more difficult to analyze and to draw causal conclusions from because of the potential for biases due to selection, confounding, and reverse causation.

Observational studies are used to characterize populations, their changes over time, to identify the risks of common exposures, to identify or test predictors or causes of health outcomes, and to describe the natural history of diseases. They may also be used to test the rare or long-term effects of treatments that are not practical to explore in clinical trials.

Common observational study designs include the following:

Cross-sectional studies rely on recruiting a representative cohort of participants and measuring them once. These can be used for describing the distribution of microbiome features in a population and for estimating associations between microbiomes and other factors. However, the lack of a temporal element means these studies are difficult to use to infer causal effects from.

Case-control studies can quickly identify associations between microbial features and disease outcomes, by comparing exposures to potential risk factors among "cases" selected from individuals with a specific outcome to "controls" representing the general population. The main advantage over a cross-sectional study for microbiome research is that outcomes that would be rare in a population representative cross-section are selected for in the case-control design and so are more easily studied. Difficulties include the identification of cases and of control groups that are representative of the rate of exposure in the population from which the cases arise [31]. Identifying causal effects is also a challenge in microbiome-based case-control studies, because in contrast to case-control studies in many other areas, the exposure of interest (usually the microbiome) cannot be assessed retrospectively and so must be measured contemporaneously or even after the occurrence of the outcome.

The control selection is complex and requires careful consideration as the way in which controls are selected has important implications for data analysis and interpretation of effects [32]. If cases are rare or difficult to recruit, precision can be improved by including more than one control per case [31].

If cases and controls differ on potentially confounding characteristics, then statistical control of these differences is important. Matching in case-control studies enables this confounding control without loss of efficiency. Matching can be done on the individual level (matching each case to one or more corresponding controls) or at the cohort level.

Matched designs cannot be used to study the effect of a matching factor on the outcome, and can bias results if controls are made too similar to cases with respect to the exposure being tested as well as confounding factors [33]. For example, matching cases and controls within households may obscure microbiome-related effects on outcomes, since the microbiomes of household control could be more like those of the cases than those of the general population would be, biasing estimates of differences between groups.

Note that matching does not by itself remove confounding effects associated with the matching factors, statistical control in the analysis must still include all of the factors on which the cohorts are matched, or an indicator representing the matched groups [34].

Longitudinal (prospective cohort) studies: Here, participants are followed prospectively, usually to observe a change in the microbiome over time, or to monitor the influence of microbial features measured at baseline on subsequent health outcomes. These provide clearer estimates of causal effects than case-control studies, since temporal relationships can be directly observed, but take more time, are more costly to conduct, and suffer from attrition.

Participants for longitudinal cohort studies might be selected at random from a local population or specific group of individuals, or in matched cohorts stratified on, for example, disease status. In contrast to case-control studies which are selected on outcome, they are selected only on baseline status and so enable a wide range of outcomes to be compared between groups but can require large sample sizes and long follow-up periods for rare health outcomes to become apparent.

When designing a prospective cohort study, it is important to identify whether the goals are primarily causal or predictive, as this has implications for design decisions. Causal studies focus on understanding mechanisms, with implications for policies around modifying risk factors for prevention, while diagnostic or predictive studies focus on identifying biomarkers for diagnosis or risk prediction. Cohort studies with a causal focus should ensure that sufficient data is collected to enable methods for identification of causal effects, that is, the risk of spurious associations from reverse causation or potentially confounding factors can be removed as well as possible, either using regression models or more advanced epidemiologic methods such as Mendelian randomization.

For diagnostic or predictive studies, identification of causal effects is less important, but since the population and the setting can have a substantial effect on predictive performance metrics, attention should be paid to ensure that the sample is representative of the population to which the study findings will be applied and that outcome measures are defined to the appropriate clinical standards. Checklists such as STARD or TRIPOD should be used to anticipate the reporting requirements.

2.3 Calculate the Required Sample Size

Choosing the number of experimental units (i.e., the "sample size") is an important part of any research study design. The larger the sample size, the more precise the estimates from the study will be, and the minimum detectable effect sizes will be smaller. Underpowered studies can also lead to higher false positive rates [35].

The multidimensional nature of microbiomes complicates sample size calculations, but comprehensive guides are available to support sample size calculations for different research questions [36]. Although some microbiome-specific tools exist [37, 38], general power and sample size tools can often be applied successfully [39], as well as ideas from studies based on other types of multidimensional genomic datasets which often have a similar structure and aims. Ideally, a formal sample size calculation should be undertaken to ensure that the goals of the study are likely to be met, using the following process:

1. Identify the key research questions and the statistical methods to be used to address them.
2. Choose the required power to detect specific effects, or precision for the effect estimates.
3. Specify plausible effect sizes that the study, at a minimum, should be designed to be able to detect.
4. Identify a source of data to estimate the required variance parameters.
5. Identify a method, either analytical or by simulation, to calculate the expected power of the study for a given number of complete data points.
6. Adjust the required sample size for likely data loss.

Any justifiable method for calculating sample sizes is acceptable, but for microbiome studies, three broad approaches are typically applied, depending on the goals of the study:

1. For studies targeting specific univariate features of the microbiome, for example, alpha diversity, antimicrobial gene load, the abundance of a specific taxon or a specific ratio of abundances, standard approaches to sample size calculations can be adopted. The statistical analysis for a univariate outcome might be a one-way ANOVA, paired or unpaired *t* test, more complex linear model, or nonparametric test depending on the design, and each of these has well-established and commonly available sample size calculation procedures available. This approach is illustrated for the comparison of alpha diversity in a case-control study of two clinical phenotypes in [40].
2. For exploratory studies where the aim is to identify which microbial taxa are differentially abundant between groups, a similar analysis might be performed as for the univariate

outcome, but with additional consideration for the large number of hypothesis tests and hence the multiplicity correction that will be needed [36]. In this case, the power can be thought of as the proportion of true effects that should be detected, while controlling the false-positive rate.

3. For studies where estimating overall differences in microbiomes is of primary interest, sample size can be calculated for comparisons of beta-diversity distribution between and within groups, using for example, PERMANOVA, although sample size calculations can be based on simpler analytical techniques such as ANOVA, if the anticipated between- and within-group beta-diversity distributions are known [see for example the approach outlined in 40, 41]. Power can be calculated by simulation based on PERMANOVA analyses directly, but this relies on a method to simulate realistic datasets with different interpretable effect sizes between groups. A method to achieve this based on the coefficient of determination (ω^2) is enabled by the *micropower* R package [illustrated by 37]. The IMPACTT investigators provide a thorough discussion of these approaches, with guidance on obtaining realistic parameters on which to base calculations [42].

Calculated sample size corresponds to the data included in the final analysis and so should be adjusted for participant dropout, failures of sample processing, or data point removal at the analysis phase. Depending on the study, the required sample size should be inflated by 10–30%. Those conducted in population representatives rather than volunteer cohorts, among older, less healthy individuals, or those with longer follow-up, should anticipate higher drop-out rates.

In practice, sample sizes are often limited by resource constraints and the availability of participants and samples. An underpowered study limited by resources might still be worth conducting, potentially as a pilot study, or if the research area is new and sufficiently important, and if even imprecise estimates of effects would be of value. It might be however that a sample size calculation forces a reevaluation of which study aims are feasible, or if a study should go ahead at all.

Note that simple sample size calculations may not directly apply to complex designs with repeated measures or clustered randomizations. These designs will require further adjustments for intraclass correlations and design effects.

2.4 Select Outcomes and Measures

2.4.1 Measuring the Gut Microbiome

Fecal microbiomes are the most used proxies for the intestinal, or gut, microbiome because of their relative ease of collection, but researchers should be mindful of their limitations and consider alternative approaches where necessary and feasible.

In particular, the composition and activity of the microbiome present within a fecal sample may not represent accurately the microbiome within the gut [43, 44], which also varies considerably with location along and within the gastrointestinal tract and the mucosal layers [45]. The fecal microbiome might also be affected by diet, medications, stool consistency, and the process of excretion and later sample collection and storage. To mitigate potential biases from these sources, ensure randomization of participants to treatments in experimental studies, standardization of sample collection processes as far as practicable, and measurement and statistical control of aspects of sample collection that cannot be completely standardized, participant diet, timing of collection, fecal consistency, and medications [46].

More invasive sampling approaches, such as direct sampling of different regions of the gut via biopsy, intubation of the small intestine, or smart capsules from noninvasive sampling along the gut, are possible and should be explored where feasible or when specific effects in different areas of the gut are of direct interest.

2.4.2 Determining Microbiome Composition and Function

The gut microbiome can be characterized in terms of its microbial composition (taxonomy), the presence and abundance of specific genes, or its functional structure. Different ways to analyze the microbiome facilitate different analyses and have different resource implications.

Selecting the right measurement method depends on the research question, sample type, resources, and the level of resolution required.

Marker gene methods 16S rRNA sequencing (for bacteria) or ITS sequencing (for fungi) are cheap but can have limited resolution at the species or strain level. For deeper functional and taxonomic insights, whole-genome shotgun metagenomics allows full-genome sequencing of all microbes, providing information on species, strain variation, and functional genes, but needs higher sequencing depth and computational resources.

When a study aims to measure the virome, it is important to consider the differences in genome size and structure compared to cellular microbes. To sequence all virus genomes, the method should include steps to sequence single-stranded and double-stranded DNA and RNA and to separate the viral fraction from the cellular fraction (see Chap. 14).

Metagenomics, viromics, and marker gene methods are limited to measuring the relative abundances of taxa and so require additional measurements of absolute microbial abundance to compare absolute abundances of taxa across samples if this is of interest.

For targeted quantification of specific microbes or functional genes, qPCR offers high sensitivity and absolute quantification but is limited to known targets and cannot distinguish its cell versus acellular and free-form sources. Alternatively, the addition of mock communities at known quantities (cells, particles, or DNA concentrations) to a sample can help calibrate measurements following sample processing and should be considered.

To assess microbial activity rather than presence, metatranscriptomics (RNA sequencing) captures gene expression in real-time. Other specialized approaches, such as metaproteomics and metabolomics (see Chap. 14), focus on microbial proteins or metabolites and are useful for studying functional outputs rather than community structure.

2.4.3 Contamination Control

Controlling, measuring, and accounting for contamination is important, particularly for low-biomass samples [47]. Positive and negative controls should be included throughout the workflow to identify sources of contamination and to mitigate their effects. Negative controls (blanks) include sample-free extraction controls and sequencing controls to identify potential contamination sources. These should be processed identical to and in parallel with actual samples and sequenced at similar depths. Positive controls, such as mock communities with known composition help validate extraction efficiency, sequencing accuracy, and bioinformatic pipelines. Specifically for human gut microbiome studies, cell-based standards are available, either from the WHO or commercial sources.

During analysis, computational approaches can identify and remove contaminant sequences by comparing their distribution patterns between samples and controls. Contamination control measures and results should be transparently documented in publications.

2.4.4 Primary, Secondary, and Exploratory Outcome Measures

Specifying hypotheses and hence primary outcomes as precisely as possible enables a more focused study and a more powerful analysis. Exploratory studies should be described as such; they are important in new research areas, but have potentially higher false positive rates, and require more strict thresholds for statistical significance to control false-positive rates, making them less powerful for the detection of effects.

Progress in defining a "healthy" microbiome enables more specific outcome measures to be selected to indicate the beneficial or harmful effect of interventions [48]. These include the abundance of specific beneficial microbiome taxa or of pathogens, aspects of diversity, and the presence of anti-microbial genes, specific metabolites, or functional pathways. However, much is still unknown as to how to characterize microbial dysbiosis [49], and anticipating and pre-specifying primary and secondary outcomes where possible does not preclude additional more exploratory analysis.

2.4.5 Health Outcomes

The primary long-term goal of microbiome research is to improve health, and if the effects of interventions on health can be demonstrated along with a mediating effect of the microbiome, then this provides much stronger direct evidence that the intervention works and stronger evidence that the microbiome itself is causally linked to health [50]. Hence, studies aimed at promoting a healthy microbiome should include a measure of health outcomes where possible, and where it is feasible that these measures might be affected within the timeframe of the proposed study.

Included health outcomes should be theoretically linked to planned interventions and their anticipated effect on or interaction with the microbiome and should be validated and recognized as biomarkers of health. Whether microbial changes or health outcomes themselves are the primary or secondary outcomes will depend on the stage of the research, initial studies may be focused on proof of concept establishing microbial changes, while later efficacy studies may focus on health outcomes, with microbiomes as a mediating or process outcome.

2.4.6 Timing of Interventions and Measurements

An important aspect of study design is determining the timing and frequency of sample collection from the study population.

Sampling intervals should be appropriate to the period over which an effect is expected or anticipated, the natural rate of change of the microbiome, and outcomes of interest. Having more assessments of the microbiome of individuals can lead to more precise estimates of effects, as well as a better understanding of the trajectory of change, since the microbiome itself is subject to measurement error and to change over time that might be unrelated to the study intervention.

High day-to-day variation in microbiome samples suggests that repeated measures even within time points can considerably increase precision in effect estimates [51]. In practice, the frequency of sample collection in longitudinal study designs is limited by logistics, resources, and subject compliance. Every sample collection in a longitudinal context increases the risk of participants dropping out of a study, and so if long-term follow-up is important, then the assessments in early parts of the study should not be too frequent. Allowing participants to miss midpoint assessments but to remain in the study and complete the most important final assessments may mitigate this problem.

2.5 Identify and Mitigate the Effects of Biological and Technical Covariates

Many different factors can affect gut microbiome composition [5, 48, 52, 53] or related outcomes, and the potential effects of each of these should be considered in study design. Uncontrolled covariates can confound the relationship between exposures or outcomes leading to spurious associations, add variance to outcome measures that lead to imprecision in the results, or affect the effectiveness of treatments.

Table 1
Covariates in microbiome studies

Category	Covariate	Mitigation strategies
Demographics	Age, sex, ethnicity, socioeconomic status	Stratification, matching, or statistical adjustment
Diet	Specific foods, fiber intake, alcohol consumption, timing of food intake	Statistical control using food diaries, standardized questionnaires, or dietary restrictions during the study period
Medications	Antibiotics, proton pump inhibitors, metformin, NSAIDs	Exclusion criteria, washout periods, statistical adjustment using medication records
Environment	Geographical location, household sharing, pet ownership, seasonal variation	Record keeping, matching, statistical adjustment
Lifestyle	Exercise, sleep patterns, stress, smoking	Validated assessment tools, statistical adjustment
Host genetics	Genetic variants affecting microbiome composition	Genetic screening, family-based designs
Technical	Sample collection method, storage conditions, DNA extraction method, sequencing platform, batch effects, date, and time of day of collection	Standardization, randomization, thorough documentation and statistical control, inclusion of technical controls

To mitigate these impacts, first identify all the factors that might affect the study outcome measures. These might relate to the participant characteristics, technical aspects of the intervention administration, data collection, or sequencing (Table 1). To mitigate the impact of each of the identified nuisance factors, one or more of the following approaches can be used:

Randomization, Blinding, and Allocation Concealment
Systematic bias can be introduced at multiple points in a microbiome study from a range of sources. Randomization should be used at all stages of a study where a systematic bias might occur. This includes the allocation of treatments not only to experimental units in interventional studies but also to aspects of the study such as sequencing, batches, and plate location for high-throughput assays or the order in which samples are processed or measured [54].

Blinding refers to the masking of group allocations to both participants and researchers and prevents investigator or participant expectations from influencing measurements and results. In randomized controlled trials, participants should be blinded to their allocations as far as possible. In all research studies, researchers should be blinded to group allocations or disease status when conducting analyses, taking measurements, or handling samples.

In experimental studies, even if blinding is not possible, it is important that the allocation sequence is concealed from those recruiting participants, to prevent selection into a study being affected by knowledge of the sequence.

Blocking and Stratification Simple randomization reduces bias but can risk an increased variance by risking an imbalance of treatment allocations. Where important nuisance factors are known (e.g., age and sex of participants, or sequencing batch), ensuring that these are balanced across experimental groups using stratified blocks can improve efficiency [54].

Homogeneity and Standardization Variation associated with a nuisance factor can be removed by holding that factor constant throughout the study. For example, restricting the participant population by inclusion or exclusion criteria, ensuring that all sequencing or other analytical work is conducted in the same batch, and that sample collection protocols are as similar as possible.

Measurement and Statistical Control Ensure that enough data is collected to enable effects of covariates on outcomes to be measured and accounted for in statistical modelling.

Care should be taken when setting out a research question to understand the role that covariates might play in the path between treatments and outcomes (e.g., confounding, colliding, mediation, or moderation), as this affects whether and how they should be included in analysis models and the interpretation of results. Creating a causal directed acyclic graph (DAG) can help understand the relationships between key variables and guide the selection of which to include in data collection and statistical models [55].

2.5.1 Metadata Collection and Management

Measuring potentially important covariates is usually achieved by questionnaires, through direct access to routinely gathered information if possible, and by carefully documenting other factors associated with the data generation process (sequencing batch numbers, dates of assessments, personnel identifiers, etc.). As with outcome measures, measurement of covariates should be made using instruments validated in your target population where possible.

While assessing potential covariates is important, do not collect more information than will be realistically used, both for ethical and practical reasons. Excessive data collection presents a risk to participants and a risk to the study if participants drop out or provide incomplete measures owing to fatigue. The questions asked should also be comprehensible, acceptable, and unambiguously answerable by the participants. Patient and public co-design and feasibility studies can help to assess this.

When nuisance factors are unavoidably present and cannot be perfectly measured and controlled in analysis, then these factors should be acknowledged in interpretation and reporting [12]. This is usually the case when observational studies are used to infer causal effects and where the possible effects of confounding factors are impossible to completely remove.

In practical terms, metadata collection can be based on established standards such as the Genomic Standards Consortium's MIxS (Minimum Information about any (x) Sequence) with the appropriate extensions [56] or the Microbiome Quality Control Project recommendations [57]. Establish systematic metadata management procedures, including standardized terminology, controlled vocabularies, and ontologies when possible. Implement quality control for metadata, including validation checks for completeness, consistency, and accuracy.

Key metadata domains to consider include the following:

1. Participant demographics (age, sex, ethnicity, etc.).
2. Clinical information (diagnosis, medications, comorbidities).
3. Environmental factors (geographical location, household characteristics).
4. Behavioral factors (diet, physical activity, smoking).
5. Sample-specific details (collection method, processing time, storage conditions).

2.6 Population Selection and Recruitment

2.6.1 Target Populations and Inclusion Criteria

Inclusion and exclusion criteria define which individuals are eligible to participate in a study.

Inclusion criteria describe the "target population" being studied, that is, the population to which the investigators will be able to directly generalize the findings. Therefore, inclusion criteria will include demographic, geographic, or clinical characteristics, the setting from which they will be recruited, and the period for recruitment of the study.

For smaller explanatory efficacy studies, where the aim is to establish a "proof of concept" for an intervention, a homogenous study group of those most likely to benefit from an intervention and most likely to complete a study may be the most appropriate choice. This will reduce variance in outcome measures and increase the likely effect size, increasing power and efficiency, but at the cost of generalizability to the wider population. A subsequent larger effectiveness study might include this wider, more heterogeneous population to estimate how well an intervention works in a more representative group [58].

There are many underserved populations in human studies, and so there should be an aim towards equity and inclusivity with regard to these groups, and reporting on gender, ethnicity, and socioeconomic background [59].

2.6.2 Exclusion Criteria

Exclusion criteria specify which, if any, subgroups of the target population should not be included in a study. Participants can be excluded for practical reasons (e.g., inability to consent or difficulty collecting samples), because they are at high risk of adverse effects from interventions, or because of a comorbidity, medication use, or lifestyle that would strongly influence outcome measures or severely modify the effect that the study is designed to test.

Table 2 lists possible exclusion criteria for microbiome studies. Consider how each criterion aligns with the specific research question and whether excluding certain groups might limit the applicability of the findings. Exclusion criteria will vary based on whether you are conducting a tightly controlled efficacy trial, or a more generalizable effectiveness trial or a population-based cohort study. Too many exclusions limit study generalizability and can prevent adequate recruitment, particularly among groups such as older people who are likely to have a lot of comorbidities and medication use, so alternative methods for control of covariates can be considered.

2.6.3 Sampling Strategy and Participant Recruitment

Once the inclusion and exclusion criteria have been defined, a plan for recruiting participants is needed.

For an observational study, random sampling should be used where possible. Typically, this means identifying a sampling frame (all those in your study population who are accessible and who might be eligible) and then inviting all or a randomly selected subset of these to be screened for participation.

Convenience or volunteer samples should be avoided in observational studies, as inferences from these samples may be biased by factors related to propensity to join the sample, leading to false positive associations arising directly from the selection and recruitment process [55].

It is important that the process of selection and recruitment of individual participants is completely documented, including those who are invited but not recruited, so that generalizability to the target population and beyond can be evaluated, and sampling weights can be applied in analysis where necessary.

Random selection of participants is not always possible in observational studies, for example, when using samples from a biorepository or other existing study, but the potential for selection bias in these cases should be noted as a study limitation.

Randomized controlled trials use random allocation of participants to treatments for the validity of their inferences, and so it is less important that the study sample itself is representative of the sample population. Hence, random selection of patients from the population is not usually required; convenience and volunteer samples are acceptable and are typically used for RCTs. Heterogeneity of treatment effects across groups might be a concern, in which case

Table 2
Potential exclusion criteria in microbiome studies

Exclusion criterion	Justification	Considerations
Recent antibiotic use (typically within 1–6 months)	Antibiotics profoundly alter microbiome composition, potentially masking or modifying effects	Consider recording rather than excluding if studying populations with high antibiotic use (e.g., older adults, hospitalized patients); may specify classes of antibiotics
Use of medications affecting gut function (PPIs, metformin, laxatives)	These medications significantly alter gut pH, transit time, and microbiome composition	Consider recording medication use and including as covariates rather than excluding if prevalent in the population of interest
Use of probiotics or prebiotics (typically within 1–3 months)	These supplements directly modify gut microbiota	Duration of exclusion should reflect the specific product and its expected washout period
Gastrointestinal disorders (IBD, IBS, celiac disease, recent gastrointestinal surgery)	These conditions are associated with altered microbiome composition, gastrointestinal environment, or transit time	May be inappropriate to exclude if studying populations where these conditions are common; consider stratification instead
Major dietary restrictions or special diets	Diet strongly influences microbiome composition	Consider recording detailed dietary information rather than excluding
Severe comorbidities (e.g., cancer, liver/kidney disease)	These conditions and their treatments can alter microbiome composition	Consider the prevalence of these conditions in your target population; overly strict criteria can limit generalizability
Recent travel to regions with different endemic microbiota	Travel can introduce temporary shifts in gut microbiota	Define "recent" based on expected stabilization period (typically 2–4 weeks)
Inability to provide informed consent	Ethical requirement for research participation	Consider if legally authorized representatives can provide consent if the population of interest includes those with cognitive impairment
Pregnancy or breastfeeding	Hormonal changes and ethical considerations regarding research risks	Important exclusion for intervention studies
Alcohol abuse or illicit drug use	These can significantly alter gut microbiota and may introduce compliance issues	Consider recording consumption patterns rather than excluding if studying populations where substance use is common
Participation in other clinical trials	Potential for conflicting interventions or assessment burden	Specify time frame (e.g., concurrent trials or within the last 30 days)
Poor compliance with sampling procedures	Compromises data quality and study validity	Consider assessing during a run-in period rather than after enrolment

targeting recruitment across subpopulations followed by blocked randomization and subgroup analysis is possible.

2.6.4 Participant Engagement and Retention

Microbiome studies in humans require the active engagement of study participants for longitudinal studies that will require multiple sample collections over a long period of time. A considerable literature [60] suggests that a well-planned engagement strategy can help improve participant compliance and reduce dropout rates. This engagement strategy should include a consistent and transparent communication with the participants, such as providing clear written instructions for sample collection and storage/transport. Additionally, consider developing an understandable educational resource about the study's purpose and importance and provide regular study updates to maintain engagement.

To promote compliance, make the sample collection as convenient as possible and consider the participant burden when designing sampling schedules and assessments of outcomes. Implementing tracking systems to monitor participation can help identify compliance issues early and mitigate them. If long-term outcomes are important, then remind participants to regularly update contact details so that they are not lost to follow-up.

Strategies to reduce and manage missing data should be considered in design and analysis and included in protocols submitted for ethical review. It is important to document reasons for missed samples or study withdrawal and to establish procedures for following up on missed appointments or samples.

Early identification of compliance issues can allow for intervention before data quality is compromised. However, all engagement strategies should be implemented in accordance with ethical guidelines and should not create undue pressure on participants.

Pilot and feasibility studies can help to identify potential issues with processes, and these can be anticipated by discussion with representatives of participant populations through patient and participant involvement activities.

3 Notes

Researchers interested in human gut microbiome science should understand both the principles of good clinical study design and best practices with sampling and analyzing microbiomes. A lot of guidance and reporting checklists are available to support the development of standardized high-quality protocols and research designs. While many methodological challenges remain in understanding the complex relationships between microbiomes and health outcomes, careful study design provides the essential foundation for advancing the field. We encourage researchers to engage with the vast and rapidly evolving literature to select the best methods to answer novel and important questions and contribute to the establishment of best practices in microbiome research.

Acknowledgements

This work was funded by Biotechnology and Biological Sciences Research Council (BBSRC) through an Institute Strategic Programme award to the QIB Programmes Gut Microbes and Health BB/R012490/1 and their constituent project(s), and associated Core Capability Grant BB/CCG1860/1 and Food, Microbiome and Health [BB/X011054/1] and their constituent project (s), and associated Core Capability Grant BB/CCG2260/1, and Microbes and Food Safety BB/X011011/1 and its constituent projects BBS/E/F/000PR13634, BBS/E/F/000PR13635, and BBS/E/F/000PR1363.

References

1. Hulley SB, Cummings SR, Browner WS et al (2011) Designing clinical research. Lippincott Williams & Wilkins
2. Rothman KJ, Lash TL, VanderWeele TJ, Haneuse S (2021) Modern epidemiology, 4th edn. Wolters Kluwer / Lippincott Williams & Wilkins, Philadelphia
3. Sterrantino AF (2024) Observational studies: practical tips for avoiding common statistical pitfalls. Lancet Reg Health – Southeast Asia 25:100415. https://doi.org/10.1016/j.lansea.2024.100415
4. Diacova T, Cifelli CJ, Davis CD et al (2025) Best practices and considerations for conducting research on diet–gut microbiome interactions and their impact on health in adult populations: an umbrella review. Adv Nutr 16:100419. https://doi.org/10.1016/j.advnut.2025.100419
5. Swann JR, Rajilic-Stojanovic M, Salonen A et al (2020) Considerations for the design and conduct of human gut microbiota intervention studies relating to foods. Eur J Nutr 59: 3347–3368. https://doi.org/10.1007/s00394-020-02232-1
6. Zaura E, Pappalardo VY, Buijs MJ et al (2000) (2021) optimizing the quality of clinical studies on oral microbiome: a practical guide for planning, performing, and reporting. Periodontol 85:210–236. https://doi.org/10.1111/prd.12359
7. Goodrich JK, Di Rienzi SC, Poole AC et al (2014) Conducting a microbiome study. Cell 158:250–262. https://doi.org/10.1016/j.cell.2014.06.037
8. Protocol – Health Research Authority. https://www.hra.nhs.uk/planning-and-improving-research/research-planning/protocol/. Accessed 20 Jun 2025
9. Good Clinical Practice (GCP) | NIHR. https://www.nihr.ac.uk/career-development/clinical-research-courses-and-support/good-clinical-practice. Accessed 20 Jun 2025
10. Teresi JA, Yu X, Stewart AL, Hays RD (2022) Guidelines for designing and evaluating feasibility pilot studies. Med Care 60:95–103. https://doi.org/10.1097/MLR.0000000000001664
11. UK Research Integrity Office (2023) Recommended checklist for researchers. UK Research Integrity Office
12. Vandenbroucke JP, Von Elm E, Altman DG et al (2014) Strengthening the reporting of observational studies in epidemiology (STROBE): explanation and elaboration. Int J Surg 12:1500–1524. https://doi.org/10.1016/j.ijsu.2014.07.014
13. Moher D, Jones A, Lepage L, For The Consort Group (2001) Use of the CONSORT statement and quality of reports of randomized trials: a comparative before-and-after evaluation. JAMA 285:1992. https://doi.org/10.1001/jama.285.15.1992
14. Mirzayi C, Renson A, Genomic Standards Consortium et al (2021) Reporting guidelines for human microbiome research: the STORMS

checklist. Nat Med 27:1885–1892. https://doi.org/10.1038/s41591-021-01552-x

15. Bharucha T, Oeser C, Balloux F et al (2020) STROBE-metagenomics: a STROBE extension statement to guide the reporting of metagenomics studies. Lancet Infect Dis 20:e251–e260. https://doi.org/10.1016/S1473-3099(20)30199-7
16. Simera I, Moher D, Hirst A et al (2010) Transparent and accurate reporting increases reliability, utility, and impact of your research: reporting guidelines and the EQUATOR network. BMC Med 8:24. https://doi.org/10.1186/1741-7015-8-24
17. Haynes RB (2006) Forming research questions. J Clin Epidemiol 59:881–886. https://doi.org/10.1016/j.jclinepi.2006.06.006
18. Chang SM, Matchar DB, Smetana GW, Umscheid CA (2012) Methods guide for medical test reviews. Agency for Healthcare Research and Quality (US), Rockville
19. Van Tulder M, Furlan A, Bombardier C, Bouter L (2003) Updated method guidelines for systematic reviews in the Cochrane collaboration Back review group. Spine 28:1290–1299. https://doi.org/10.1097/01.BRS.0000065484.95996.AF
20. Duvallet C (2018) Meta-analysis generates and prioritizes hypotheses for translational microbiome research. Microb Biotechnol 11:273–276. https://doi.org/10.1111/1751-7915.13047
21. Glass DJ (2014) Experimental design for biologists, 2nd edn. Cold Spring Harbor Laboratory Press, Cold Spring Harbor
22. Lazic SE (2016) Experimental Design for Laboratory Biologists: Maximising information and improving reproducibility, 1st edn. Cambridge University Press
23. Ruxton GD, Colegrave N (2016) Experimental design for the life sciences, 4th ed. Oxford university press, Oxford
24. Senn S (2021) Statistical issues in drug development, 3rd edn. Wiley Blackwell, Hoboken
25. Celentano DD, Szklo M, Gordis L (2019) Gordis epidemiology, 6th edn. Elsevier, Philadelphia
26. Berger VW, Bour LJ, Carter K et al (2021) A roadmap to using randomization in clinical trials. BMC Med Res Methodol 21:168. https://doi.org/10.1186/s12874-021-01303-z
27. Hemming K, Taljaard M (2023) Key considerations for designing, conducting and analysing a cluster randomized trial. Int J Epidemiol 52:1648–1658. https://doi.org/10.1093/ije/dyad064
28. Baker TB, Smith SS, Bolt DM et al (2017) Implementing clinical research using factorial designs: a primer. Behav Ther 48:567–580. https://doi.org/10.1016/j.beth.2016.12.005
29. Leeming ER, Johnson AJ, Spector TD, Le Roy CI (2019) Effect of diet on the gut microbiota: rethinking intervention duration. Nutrients 11:2862. https://doi.org/10.3390/nu11122862
30. Johnson AJ, Zheng JJ, Kang JW et al (2020) A guide to diet-microbiome study design. Front Nutr 7:79. https://doi.org/10.3389/fnut.2020.00079
31. Grimes DA, Schulz KF (2005) Compared to what? Finding controls for case-control studies. Lancet 365:1429–1433. https://doi.org/10.1016/S0140-6736(05)66379-9
32. Iwagami M, Shinozaki T (2022) Introduction to matching in case-control and cohort studies. Ann Clin Epidemiol 4:33–40. https://doi.org/10.37737/ace.22005
33. Bloom MS, Schisterman EF, Hediger ML (2007) The use and misuse of matching in case-control studies: the example of PCOS. Fertil Steril 88:707–710. https://doi.org/10.1016/j.fertnstert.2006.11.125
34. Pearce N (2016) Analysis of matched case-control studies. BMJ 352:i969 https://doi.org/10.1136/bmj.i969
35. Ioannidis JPA (2005) Why Most published research findings are false. PLoS Med 2:e124. https://doi.org/10.1371/journal.pmed.0020124
36. Ferdous T, Jiang L, Dinu I et al (2022) The rise to power of the microbiome: power and sample size calculation for microbiome studies. Mucosal Immunol 15:1060–1070. https://doi.org/10.1038/s41385-022-00548-1
37. Kelly BJ, Gross R, Bittinger K et al (2015) Power and sample-size estimation for microbiome studies using pairwise distances and PERMANOVA. Bioinforma Oxf Engl 31:2461–2468. https://doi.org/10.1093/bioinformatics/btv183
38. Chen L (2020) Powmic: an R package for power assessment in microbiome case-control studies. Bioinforma Oxf Engl 36:3563–3565. https://doi.org/10.1093/bioinformatics/btaa197
39. La Rosa PS, Brooks JP, Deych E et al (2012) Hypothesis testing and power calculations for taxonomic-based human microbiome data. PLoS One 7:e52078. https://doi.org/10.1371/journal.pone.0052078
40. Casals-Pascual C, González A, Vázquez-Baeza Y et al (2020) Microbial diversity in clinical

microbiome studies: sample size and statistical power considerations. Gastroenterology 158: 1524–1528. https://doi.org/10.1053/j.gastro.2019.11.305

41. Gail MH, Wan Y, Shi J (2021) Power of microbiome Beta-diversity analyses based on standard reference samples. Am J Epidemiol 190: 439–447. https://doi.org/10.1093/aje/kwaa204
42. IMPACTT Investigators (2022) Beta-diversity distance matrices for microbiome sample size and power calculations – How to obtain good estimates. Comput Struct Biotechnol J 20: 2259–2267. https://doi.org/10.1016/j.csbj.2022.04.032
43. Ahn J-S, Lkhagva E, Jung S et al (2023) Fecal microbiome does not represent whole gut microbiome. Cell Microbiol 2023:6868417. https://doi.org/10.1155/2023/6868417
44. Levitan O, Ma L, Giovannelli D et al (2023) The gut microbiome–does stool represent right? Heliyon 9:e13602. https://doi.org/10.1016/j.heliyon.2023.e13602
45. Donaldson GP, Lee SM, Mazmanian SK (2016) Gut biogeography of the bacterial microbiota. Nat Rev Microbiol 14:20–32. https://doi.org/10.1038/nrmicro3552
46. Vandeputte D, Falony G, Vieira-Silva S et al (2016) Stool consistency is strongly associated with gut microbiota richness and composition, enterotypes and bacterial growth rates. Gut 65: 57-62 https://doi.org/10.1136/gutjnl-2015-309618
47. Fierer N, Leung PM, Lappan R et al (2025) Guidelines for preventing and reporting contamination in low-biomass microbiome studies. Nat Microbiol 10:1570–1580. https://doi.org/10.1038/s41564-025-02035-2
48. Van Hul M, Cani PD, Petitfils C et al (2024) What defines a healthy gut microbiome? Gut 73:1893-1908 https://doi.org/10.1136/gutjnl-2024-333378
49. Brüssow H (2020) Problems with the concept of gut microbiota dysbiosis. Microb Biotechnol 13:423–434. https://doi.org/10.1111/1751-7915.13479
50. Hutkins R, Walter J, Gibson GR et al (2025) Classifying compounds as prebiotics — scientific perspectives and recommendations. Nat Rev Gastroenterol Hepatol 22:54–70. https://doi.org/10.1038/s41575-024-00981-6
51. Vandeputte D, De Commer L, Tito RY et al (2021) Temporal variability in quantitative human gut microbiome profiles and implications for clinical research. Nat Commun 12: 6740. https://doi.org/10.1038/s41467-021-27098-7
52. Vujkovic-Cvijin I, Sklar J, Jiang L et al (2020) Host variables confound gut microbiota studies of human disease. Nature 587:448–454. https://doi.org/10.1038/s41586-020-2881-9
53. Falony G, Vandeputte D, Caenepeel C et al (2019) The human microbiome in health and disease: hype or hope. Acta Clin Belg 74:53–64. https://doi.org/10.1080/17843286.2019.1583782
54. Burger B, Vaudel M, Barsnes H (2021) Importance of block randomization when designing proteomics experiments. J Proteome Res 20: 122–128. https://doi.org/10.1021/acs.jproteome.0c00536
55. Lipsky AM, Greenland S (2022) Causal directed acyclic graphs. JAMA 327:1083–1084. https://doi.org/10.1001/jama.2022.1816
56. Yilmaz P, Kottmann R, Field D et al (2011) Minimum information about a marker gene sequence (MIMARKS) and minimum information about any (x) sequence (MIxS) specifications. Nat Biotechnol 29:415–420. https://doi.org/10.1038/nbt.1823
57. Sinha R, Abnet CC, White O et al (2015) The microbiome quality control project: baseline study design and future directions. Genome Biol 16:276. https://doi.org/10.1186/s13059-015-0841-8
58. Le-Rademacher J, Gunn H, Yao X, Schaid DJ (2023) Clinical trials overview: from explanatory to pragmatic clinical trials. Mayo Clin Proc 98:1241–1253. https://doi.org/10.1016/j.mayocp.2023.04.013
59. Bodicoat DH, Routen AC, Willis A et al (2021) Promoting inclusion in clinical trials—a rapid review of the literature and recommendations for action. Trials 22:880. https://doi.org/10.1186/s13063-021-05849-7
60. Wong CA, Song WB, Jiao M et al (2021) Strategies for research participant engagement: a synthetic review and conceptual framework. Clin Trials Lond Engl 18:457–465. https://doi.org/10.1177/17407745211011068

Chapter 3

Collection, Storing, and Processing of Fecal Samples for Human Microbiome Research

Sarah Phillips, Rachel Watt, and Sarah Hughes

Abstract

Fecal samples are commonly used in human microbiome research, offering noninvasive access to the intestinal microbial ecosystem and the investigation of its contributions to health. This chapter discusses different sample collection, storage, and initial processing methods, emphasizing the importance of maintaining sample integrity to ensure reliable and reproducible results.

Key words Fecal, Stool, Sample collection, Microbiome, Sampling, Microbiome stabilization, Microbiome sample storage, Microbiome sample transport

1 Introduction

Fecal (stool) samples are often employed in human microbiome research as a proxy for investigating the gut microbiome. They provide a noninvasive and cost-effective way of collecting samples that can reflect distal colon microbial diversity [1, 2].

Sample collection strategies for stool samples significantly impact the viability and diversity of the microbiota recoverable in the laboratory particularly for oxygen-sensitive (obligate) anaerobes leading to shifts in microbial composition that impact recovery of viable microbes and accurate profiling of microbial communities [3–7].

Practical aspects of sample collection are also important considerations. While clinical studies benefit from being able to process samples immediately upon receipt, many microbiome studies rely on at-home self-collection kits which introduces logistical challenges, including variable storage conditions, transportation delays, and participant compliance issues [8]. Cost is another major consideration particularly for large-scale and longitudinal research projects. High-quality preservation methods, such as immediate freezing at −80 °C or the use of specialized stabilizing reagents,

Simon R. Carding (ed.), *Best Practice in Microbiome Research*, Springer Protocols Handbooks, https://doi.org/10.1007/978-1-0716-5009-7_3,

may not be financially feasible for studies collecting thousands of samples over time or those based in remote/field settings. Finally, well-established preservation methods, such as freezing, can also introduce biases in microbial composition with freezing altering the abundance of certain microbial groups, with even brief exposure to ambient temperatures can lead to compositional shifts [9]. This chapter provides three examples of fecal sample collection kits and procedures for human microbiome studies adapted from methods used at the Quadram Institute.

2 Materials

2.1 Essential Equipment

1. Ultra-low temperature (ULT) freezer capable of reaching temperatures of −80 °C.

2.2 Fecal Sample Collection Kit Method 1

1. Participant sample collection instructions (*see* **Note 1**).
2. Sample collection sticker/log (*see* **Note 2**).
3. Sample collection pot and frame (Specimen Collect Commode Grad 1Qt/.95 L, Medical Action Industries, or similar) (*see* **Note 3**).
4. Small commode liner (without absorbent material) (*see* **Note 4**).
5. 95-kPa medical specimen transport bags C3 size 305 × 403 mm for transport of category B biological samples to UN3373 P650 packaging instructions (SecureLab 95—ADR Specimen Transport Bags, or similar) (*see* **Note 5**).
6. Absorbent sheet for mailing of UN3373 Category B Biological Samples (Alpha Laboratories Absorbent Sheet 250 × 300 mm, or similar).
7. Small cool box and ice packs (Coleman 4.7 Litre Cool Box, or similar) (*see* **Note 6**).
8. Oxoid™ AnaeroGen 2.5 ltr.

2.3 Fecal Sample Collection Kit Method 2

1. Participant sample collection instructions (*see* **Note 1**).
2. Sample collection sticker/log (*see* **Note 2**).
3. Sample collection device (FecesCatcher Tag Hemi VOF, or similar) (*see* **Note 7**).
4. 20 mL universal container with spoon.
5. 95-kPa medical specimen transport bags C3 size 305 × 403 mm for transport of category B biological samples to UN3373 P650 packaging instructions (SecureLab 95—ADR Specimen Transport Bags, or similar)

6. Absorbent sheet for mailing of UN3373 Category B Biological Samples (Alpha Laboratories Absorbent Sheet 250 × 300 mm, or similar).
7. Benchtop freezer (*see* **Note 8**).
8. Cool box and ice packs (30 Litre Thermal Carry Bag With Trolley, Corrmed, or similar).

2.4 Fecal Sample Collection Kit Method 3

1. Participant sample collection instructions (*see* **Note 1**).
2. Sample collection sticker/log (*see* **Note 2**).
3. Sample collection device (FecesCatcher Tag Hemi VOF, or similar) (*see* **Note 7**).
4. OMNIgene™•GUT (OM-200).
5. Secure all-in-one package for posting specimens and samples (*see* **Note 9**).

2.5 Sample Processing Materials for all Methods

1. Sample receipt/storage log (*see* **Note 10**).
2. Cryogenic labels (capable of withstanding long-term storage at −80 °C).
3. 2 mL cryovials, sterile, and DNAse-free, externally threaded with a washer.
4. Sterile microbiological inoculation loops (*see* **Note 11**) (Microspec™ Sterile Plastic Inoculation Loops (volume: 10 μL), or similar).
5. 10% glycerol solution (made from sterile 100% glycerol and sterile nuclease-free water).
6. 3 mL Pasteur pipettes, sterile, and DNAse-free (*see* **Note 2**)
7. Wooden tongue depressor/spatula.
8. Precision weighing scales with a resolution of 0.0001 g.
9. Microbiological safety cabinet.
10. −80 freezer with appropriate racking
11. Cryobox compatible with the cryovials selected (Simport Cryostore Box, Series 281, Green Lid, for 81× 1–2 mL Cryotubes, or similar).

3 Methods

3.1 Fecal Sample Collection Method 1

This collection kit is designed to collect an intact stool sample, maintaining the sample at 0 °C or lower for 24 h in an atmosphere with reduced oxygen content for sequence-based metagenomic and culture-based analyses (*see* **Note 12**).

1. Provide the participant with all sample collection kit materials.
2. Instruct the participant to collect their sample at home according to the sample collection instructions in their kit which involves the following:
   - i. Donors must freeze the provided icepacks at least 12 h before intending to collect a sample.
   - ii. Before collecting a sample, donors should empty their bladder, so urine does not contaminate the sample.
   - iii. Donors should remove the lid of the provided commode and put the liner bag inside.
   - iv. Using the commode frame provided, donors should position the commode at the back of the toilet.
   - v. Donors should then collect a sample into the commode liner, being careful not to contaminate it with toilet paper or urine.
   - vi. Donors should then remove the commode from the toilet, fold over the liner (but not seal the liner shut) and place the opened AnaeroGen sachet inside the pot but on the outside of the commode liner, and then seal the commode pot tightly with the lid.
   - vii. Donors should write the date and time that the sample was collected on the label on the lid of the commode.
   - viii. Donors should place the sealed commode and the white absorbent material into the large Medical Specimen Transport Bag and seal by removing the silver strip and closing.
   - ix. The sample should then be transferred into the cool box with the frozen icepacks.
3. Donors should bring their sample in person to the laboratory as soon as possible after collection and within 24 h (*see* **Note 13**) of production (a courier service can also be arranged to collect the sample from the donor).
4. Upon receipt in the laboratory, transfer the cool box to a microbiological safety cabinet.
5. Remove the specimen from the cool box and outer packaging.
6. Record the time and date of receipt, as well as any quality or packing issues (for instance, high temperature) on the sample receipt log.
7. Homogenize the sample within the bag by stirring each stool sample thoroughly using a spatula or wooden tongue depressor until fully homogenized: the color, consistency, and content of the whole stool sample must look and/or feel the same (*see* **Note 14**).

8. Label the required number of cryovials for storage/further processing of the sample according to local policy (*see* **Note 15**).
9. Aliquot the sample into the cryovials, noting the weight of the sample transferred to the vial. Depending on the required onward processing and duration of storage required, 10% glycerol can be added as a cryoprotectant for long-term storage (*see* **Notes 16** and **17**).
10. Transfer the cryovials into a cryobox in the ULT (−80 °C) freezer until required (*see* **Note 18**) noting the position of the sample within the cryobox and the location of the cryobox within the freezer on the sample receipt log (or sample management system if available).

3.2 Fecal Sample Collection Method 2

This collection kit is designed to collect a part of a stool sample that is frozen immediately in a home freezer (*see* **Note 19**) to preserve the microbiome profile for sequence-based metagenomic and culture-based analyses (*see* **Note 8**). This method has been used in a study observing microbiome composition in a cohort of pregnant women throughout pregnancy, at birth, and for the first 2 years of the infant's life.

1. Provide the participant with all sample collection kit materials and arrange for a benchtop freezer to be delivered and set up at the participant's home.
2. Instruct the participant to collect their sample at home according to the sample collection instructions in their kit which involves the following:
   (i) Before collecting a sample, donors should empty their bladder, so urine does not contaminate the sample.
   (ii) To ensure secure adhesion of the sample collection device (FecesCatcher), donors should dry the toilet seat with toilet paper before sticking the device to the rim of the toilet following the instructions provided.
   (iii) A stool sample is collected in the FecesCatcher ensuring that used toilet paper does not touch the stool sample.
   (iv) The stool sample is deposited into the collection pot until approximately half full and then securely sealed, writing the time and date of sample collection on the sample collection pot.
   (v) The FecesCatcher is fully biodegradable and can be flushed with any remaining stool once the sample has been collected.
   (vi) The sample is then immediately transferred to the provided benchtop freezer prior to collection.

3. Once the donor has notified the research team that their sample has been collected, the sample is collected by the research team or courier using cold chain storage (such as with a cool box and icepacks) to maintain a temperature of 0 °C or lower. A digital temperature monitor can be employed to monitor and record conditions during transit (*see* **Note 13**).
4. Upon receipt in the laboratory, transfer the cool box to a microbiological safety cabinet.
5. Remove the specimen from the cool box and outer packaging.
6. Record the time and date of sample collection as well as sample receipt, as well as any quality or packing issues (e.g., high temperature) on the sample receipt log.
7. Label the required number of cryovials for storage/further processing of the sample according to local policy (*see* **Note 15**).
8. Aliquot the sample into the cryovials, noting the weight of the sample transferred on the vial.
9. Depending on the required onward processing and duration of storage required, 10% glycerol can be added as a cryoprotectant for long-term storage (*see* **Note 17**).
10. Transfer the aliquots into a cryobox into the -80 °C freezer until required (*see* **Notes 16** and **18**) noting the position of the sample within the cryobox and the location of the cryobox within the freezer on the sample receipt log (or sample management system if available).

3.3 Sample Collection Kit 3 Method

This collection kit is designed to collect a small sample of stool that is preserved in DNA stabilization buffer (*see* **Note 20**). Sample collection is remote and relatively low cost compared to other methods which is advantageous for studies with large numbers of participants, or if participants are spaced widely geographically making it impractical to deliver their sample directly to the laboratory. The method can be used for studies conducting genomic analysis and for which microbial viability is not a major concern.

1. Provide the participant with all sample collection kit materials.
2. Instruct the participant to collect their sample at home according to the instructions provided with the sample collection device in their kit. Samples are then mailed to the research team as soon as possible (but no longer than 2 days) after collection using the all-in-one secure postal packaging provided.
3. Upon receipt, transfer the package to a microbiological safety cabinet.
4. Remove the specimen from the box and outer packaging.

5. Record the time and date of receipt, as well as any quality or packing issues (for instance, leakage) on the sample receipt log.
6. Label the required number of cryovials for storage/further processing of the sample according to local policy (*see* **Note 15**).
7. Aliquot the sample into the cryovials, noting the weight/volume of sample transferred on the vial.
8. Transfer the aliquots into a cryobox into the -80 freezer until required (*see* **Notes 16** and **18**) noting the position of the sample within the cryobox and the location of the cryobox within the freezer on the sample receipt log (or sample management system if available).

4 Notes

1. Sample collection instructions should be as simple as possible to aid compliance. Provide clear images of the process involved with the information provided being tailored to the needs of the target audience. Before starting the study, it is advisable to seek advice and input from members of the target audience to obtain feedback as to the suitability and usability of the sample collection method and the clarity of the instructions.
2. It is important to record the date and time that samples were produced so that the time between production of the sample and processing can be recorded in the study record. This can be useful to get an indication of the time that the sample has been held under temporary conditions, and whether the protocol has been adhered to. The gut microbiome is dynamic with variances in microbiome composition seen in the same subject over the course of a single day [10] with recordings of the time at which participants have collected their sample, therefore, being informative. It is also worth asking participants to provide their samples at a similar time of day (for instance, the first motion of the day), to attempt to moderate the time of production as a potential confounder.
3. Commodes are a useful way to assist the participant in collecting their sample and avoid home remedies and solutions (e.g., takeaway pots, cling film, etc.). A rigid and sturdy commode is useful for individuals with mobility/dexterity issues as it is easier to set up than other alternatives that can require considerable manual dexterity.
4. A small commode liner should cover the pot during sample collection but be small enough that it can be folded and sealed in the commode along with the gas sachet. The Quadram Institute does not use bags that are free of DNA due to the

high cost and the likely contamination when participants are setting up the sample collection device. However, this may be worth considering for genomic analysis studies and the impact that this may have on the interpretation of results.

5. The fully packaged sample must comply with local regulations for the transport of biological material. Guidance on packing and transport requirements for patient samples in the UK can be found on the UK government website (Packaging and transport requirements for patient samples – UN3373 - GOV.UK) [11].
6. Cool box and ice packs should be big enough to contain the sample but small enough for the sample to be secured inside and in contact with the icepacks. Before use, test the ability of the cool box and the number of icepacks needed to maintain a temperature below 0 °C for 24 h (or timeframe relevant for a particular study). This can be done using a portable temperature logger as a dummy sample to record the length of time that these temperatures are held.
7. The type of sample collection device is important if fecal samples are likely to have high water content with devices such as FecesCatcher Tag Hemi VOF, or similar being less effective for such samples. If practical and affordable, extra devices could be provided as some participants have reported issues with them becoming unstuck and having to repeat the collection. Some participants have reported adding additional tape to the devices to ensure they are more firmly attached to the rim of the toilet seat.
8. Any benchtop freezer provided to study participants should be small enough for them to accommodate in their home. Energy efficiency and cost should also be a concern when sourcing the freezer. Asking participants to store stool samples in their own home freezers has health and safety implications and can be a barrier to participation [12].
9. Royal Mail Safebox™ Special Delivery is an example of a service available in the UK which is convenient for the participant as post boxes are usually near their homes.
10. It is important to record the date and time that samples arrive at the laboratory so that the time between production of the sample and the time of arrival can be recorded in the study record. This can be useful to get an indication of the time that the sample has been held under temporary conditions, and whether the protocol has been adhered to. In addition, it is important to keep a fully traceable record of what has happened to the sample from the point of receipt until final disposal of the sample. Key things to record would include (but not limited to) the following:

(i) The time and date of receipt.

(ii) The individual who accepted delivery.

(iii) How it was stored upon receipt until processed.

(iv) Characteristics of the sample and sample quality (for instance, if the sample appears to be loose or contaminated).

(v) Time and date of processing of the sample.

(vi) The individual conducting the processing.

(vii) Any reagents/kits (including batch numbers) that have been used during processing of the sample.

(viii) Serial numbers of instruments that have been used to process the sample.

(ix) Where the samples have been stored following processing.

11. Disposable DNA-free sterile microbiological loops: Less flexible loops are ideal to prevent any material from flicking off the loop when aliquoting the material which poses risks to contamination and the safety of the worker.

12. This sample collection method, while not technically demanding, has lots of steps that participants must follow and may not be suitable for participants with limited capacity such as young children or those severely unwell or hospitalized patients. Being reliant on participants to bring their sample directly into the institute also places a cost and time burden on the participants and may not be appropriate for multiple, longitudinal sample collection schedules.

13. A limitation of this method is that samples cannot be processed or frozen at −80 °C upon production of the sample. However, it has been reported that microbial composition within fecal samples can be relatively stable for up to 96 h post-collection at 4 °C [13].

14. Homogenization is important as variability occurs in microbial communities in replicate samples collected from different fractions of the same stool [14].

15. Traceability of samples and the handling condition of the samples are important throughout the duration of the study to ensure that they have been handled correctly and that the data produced has not been affected by handling conditions. Chain of custody and documentation of cold chain storage is an important consideration when designing a microbiome study. This information should be documented within the study master file. Adequate information should be added to the cryovials (ideally using a cryogenic label) so that the sample is traceable. The storage location of the sample should be recorded, and a log maintained of all samples from sample receipt to disposal.

16. Aliquoting of samples to provide enough for planned and potential future analyses is important to avoid freeze-thaw cycles which can cause substantial changes in the microbial composition of the sample [15]. The need for cryoprotective agents is another important consideration.
17. The inclusion of 10% glycerol in samples to be stored at −80 °C or in liquid nitrogen is an effective way of preserving the integrity and composition of stool samples for at least 12 months [3]. Snap freezing before storage at −80 °C should be chosen wherever possible as this reduces the impact of ice formation and its effect on the microbiome [16].
18. While processing samples immediately, or aliquoting and storing them at −80 °C, is considered the gold standard for sample storage for microbiome analysis, interim storage for 12 to 48 h at −20 °C does not significantly alter overall microbiome composition within the sample [17].
19. Research into the use of DNA stabilization buffers is equivocal, with some studies reporting significant alterations of specific phyla (Bacteroidetes, Firmicutes, and Actinobacteria) compared with samples that have been immediately frozen with no preservative, with questions remaining as to whether these variations are due to differences in susceptibility to lysis or depletion during processing of specific taxa [5].
20. To meet these aims, the sample collection kit was designed to include a DNA stabilization buffer. Several commercial collection kits have been developed by companies including OMNIgene GUT OMR-200 (DNA Genotek) and DNA/RNA Shield (Zymo Research) which yield results comparable to immediately frozen samples [15, 18–20]. The OMNIgene GUT OMR-200 (DNA Genotek) is chosen as it provides a way to measure the quantity of stool and homogenize the sample in the buffer at the point of collection.

Acknowledgements

This work was funded by the Biotechnology and Biological Sciences Research Council (BBSRC) through an Institute Strategic Programme award to the QIB Programmes Gut Microbes and Health BB/R012490/1 and Food, Microbiome and Health [BB/X011054/1] and their constituent project(s).
The sample collection protocol in Method 1 was developed and used by studies at the Quadram Institute including the MOTION study [21] led by Professor Simon R. Carding.
The sample collection protocol for Method 2 was developed and used by studies at the Quadram Institute including the BAMBI

[22], led by Professor Lindsay Hall, with additional input from other longitudinal cohort studies including the Flemish Gut Project [23].

References

1. Gill SR, Pop M, Deboy RT et al (2006) Metagenomic analysis of the human distal gut microbiome. Science 312:1355–1359. https://doi.org/10.1126/science.1124234
2. Eckburg PB, Bik EM, Bernstein CN et al (2005) Diversity of the human intestinal microbial flora. Science 308:1635–1638. https://doi.org/10.1126/science.1110591
3. Li X, Shi X, Yao Y, Shen Y, Wu X, Cai T, Liang L, Wang F (2023) Effects of stool sample preservation methods on gut microbiota biodiversity: new original data and systematic review with meta-analysis. Microbiol Spectr 11: e04297–e04222. https://doi.org/10.1128/spectrum.04297-22
4. Biclot A, Huys GRB, Bacigalupe R et al (2022) Effect of cryopreservation medium conditions on growth and isolation of gut anaerobes from human faecal samples. Microbiome 10:80. https://doi.org/10.1186/s40168-022-01267-2
5. Maghini DG, Dvorak M, Dahlen A et al (2024) Quantifying bias introduced by sample collection in relative and absolute microbiome measurements. Nat Biotechnol 42:328–338. https://doi.org/10.1038/s41587-023-01754-3
6. Isali I, Wong TR, Tian S (2024) Best practice guidelines for collecting microbiome samples in research studies, European urology. Focus 10(6):909–913., ISSN 2405-4569,. https://doi.org/10.1016/j.euf.2024.12.007
7. Nearing JT, Comeau AM, Langille MGI (2021) Identifying biases and their potential solutions in human microbiome studies. Microbiome 9:113. https://doi.org/10.1186/s40168-021-01059-0
8. Vandeputte D, Vanleeuwen R, Falony G et al (2015) Perspectives and pitfalls of microbiome research through home based fecal sampling: the Flemish gut Flora project experience. Arch Public Health 73(Suppl 1):P33. https://doi.org/10.1186/2049-3258-73-S1-P33
9. Jenkins SV, Vang KB, Gies A et al (2018) Sample storage conditions induce post-collection biases in microbiome profiles. BMC Microbiol 18:227. https://doi.org/10.1186/s12866-018-1359-5
10. Allaband C, Lingaraju A, Flores Ramos S et al (2024) Time of sample collection is critical for the replicability of microbiome analyses. Nat Metab 6:1282–1293. https://doi.org/10.1038/s42255-024-01064-1
11. GOV.UK (n.d.) Packaging and transport requirements for patient samples—UN3373. Packaging and transport requirements for patient samples – UN3373 - GOV.UK. Accessed 30 Apr 2025
12. Bolte LA, Klaassen MAY, Collij V et al (2021) Patient attitudes towards faecal sampling for gut microbiome studies and clinical care reveal positive engagement and room for improvement. PLoS One 16(4):e0249405. https://doi.org/10.1371/journal.pone.0249405
13. Holzhausen EA, Galloway-Peña J, Baral R et al (2021) Assessing the impact of storage time on the stability of stool microbiota richness, diversity, and composition. Gut Pathog 13(1):75. https://doi.org/10.1186/s13099-021-00470-0
14. Yeoh YK, Chen Z, Hui M et al (2019) Impact of inter- and intra-individual variation, sample storage and sampling fraction on human stool microbial community profiles. PeerJ 7:e6172. https://doi.org/10.7717/peerj.6172
15. Song SJ, Amir A, Metcalf JL et al (2016) Preservation methods differ in fecal microbiome stability, affecting suitability for field studies. mSystems 1(3):e00021–e00016. https://doi.org/10.1128/msystems.00021-16
16. Hickl O, Heintz-Buschart A, Trautwein-Schult A, Hercog R, Bork P, Wilmes P, Becher D (2019) Sample preservation and storage significantly impact taxonomic and functional profiles in metaproteomics studies of the human gut microbiome. Microorganisms 7(9):367. https://doi.org/10.3390/microorganisms7090367
17. Bassis CM, Moore NM, Lolans K et al (2017) Comparison of stool versus rectal swab samples and storage conditions on bacterial community profiles. BMC Microbiol 17:78. https://doi.org/10.1186/s12866-017-0983-9
18. Doukhanine E, Bouevitch A, Pozza L, Merino C (2014) OMNIgene®•GUT enables reliable collection of high-quality fecal samples for GUT microbiome studies. DNA Genotek. https://dnagenotek.com/ROW/pdf/PD-WP-00040.pdf. Accessed 30 Apr 2025

19. Doukhanine E, Bouevitch A, Pozza L, et al (2016) OMNIgene®•GUT stabilizes the microbiome profile at ambient temperature for 60 days and during transport. DNA Genotek https://www.dnagenotek.com/US/pdf/PD-WP-00042.pdf. Accessed 30 Apr 2025
20. Kazantseva J, Malv E, Kaleda A et al (2021) Optimisation of sample storage and DNA extraction for human gut microbiota studies. BMC Microbiol 21:158. https://doi.org/10.1186/s12866-021-02233-y
21. Phillips S, Watt R, Atkinson T et al (2022) A protocol paper for the MOTION study: a longitudinal study in a cohort aged 60 years and older to obtain mechanistic knowledge of the role of the gut microbiome during normal healthy ageing. PLoS One 17(11):e0276118. https://doi.org/10.1371/journal.pone.0276118
22. Alcon-Giner C, Dalby MJ, Caim S et al (2020) Microbiota supplementation with Bifidobacterium and lactobacillus modifies the preterm infant gut microbiota and metabolome: an observational study. Cell Rep Med 1(5):100077. https://doi.org/10.1016/j.xcrm.2020.100077
23. Vandeputte D, Tito RY, Vanleeuwen R et al (2017) Practical considerations for large-scale gut microbiome studies. FEMS Microbiol Rev 41(Suppl_1):S154–S167. https://doi.org/10.1093/femsre/fux027

Chapter 4

Collection, Storing, and Processing of Other Samples

Silas Triller, Gemma Beasy, Sarah Hughes, Iliana Serghiou, and Jennifer Ahn-Jarvis

Abstract

The oral cavity, skin, vagina, and colon are environments equipped with specialized structures and functions which can support complex ecosystem of microorganisms enabling them to serve as protective barriers, regulate immune responses, and contribute to metabolic processes. Samples from these sites are characterized by low biomass, contamination by human DNA, and bacteria recalcitrant to lysis. Recent advances in DNA extraction methodology and metagenomic and high-throughput sequencing technologies have improved the ability to characterize these complex and diverse microbial communities. In this chapter, sample collection methodologies for the oral cavity, skin, vagina, and biopsies from the colon are provided, including preparation of the collection site, choice of sampling tools, sample handling, transport, and storage to minimize microbial degradation and contamination.

Key words Microbiome, Colon biopsy, Low biomass, Oral mucosa, Saliva, Skin, Vagina

1 Introduction

Reproducibility and accuracy of sample collection are paramount to the quality of the downstream genomic analysis. Sampling site and tools used for their collection have a significant impact on the quality and integrity of the recovered microbial DNA. These features collectively contribute to their impact on human health and disease outcome [1]. The diversity of the human microbiome in different sites is considerable due to differences in their proximate environment, physiologic milieu, exposure to external stressors, and spatial distribution [2–8]. The samples described in this chapter have relatively low biomass with high amounts of human DNA and have intrinsic factors that degrade samples or inhibit amplification. Moreover, screening participants with well-defined inclusion and exclusion criteria plays a crucial role in ensuring accuracy and consistency in microbiome sampling [9, 10]. These criteria are important when reporting microbiome research in accordance

Simon R. Carding (ed.), *Best Practice in Microbiome Research*, Springer Protocols Handbooks,
https://doi.org/10.1007/978-1-0716-5009-7_4,

with the STORMS (Strengthening The Organization and Reporting of Microbiome Studies) guidelines [11].

1.1 Oral Sample Collection

Samples can be collected from the oral cavity including the oral mucosa (gums and buccal surfaces), teeth, tongue, and saliva with many factors influencing oral microbiome analysis [9, 12, 13]. The order in which samples are collected is an important consideration in avoiding cross-contamination of samples.

1.2 Skin Sample Collection

Skin sites can be categorized as dry, moist, and sebaceous regions, the characteristics of which are influenced by skin thickness [14], keratinization [15], hair density [16], age and gender [17], and moisture content [18]. Moreover, the use of different sampling and processing methods and storage conditions [19–21] will influence microbiome composition and microbial load.

1.3 Vaginal Sample Collection

The complex vaginal ecosystem comprises the vaginal mucosa made up of stratified squamous non-keratinized epithelium coated with cervicovaginal secretions and its colonizing microbial communities [2]. Vaginal microbiome (VMB) variations vary over a women's lifetime and according to transitional periods such as puberty, menopause, and pregnancy and according to the anatomical sampling site, sampling device and collection methods used, and storage conditions in addition to distortions related to diversity in laboratory methods, DNA extraction and processing and sequencing technology, and differing bioinformatic approaches [2, 22]. A comprehension of how these confounding factors impact VMB studies is essential to accurately interpret results and justify comparisons between studies.

1.4 Colonic Biopsy Sample Collection

Endoscopic and colonic biopsies can be collected from the ascending colon, transverse colon, descending colon, and sigmoid colon with current recommendations suggesting the collection of at least eight biopsies sampling across the whole colon [23, 24]. At the Quadram Institute, endoscopic biopsies are the most common biopsy used for microbiome analysis. Challenges include the requirement for skilled professionals and precise protocols for extracting the microbial profile from the samples, difficulties in accessing and sampling the specific sites, low biomass of the samples, and participant factors including adhering to diet restriction and bowel preparation (laxatives) prior to colonoscopy which impact microbiota composition [25]. Reductions in the amounts of DNA required for metagenomics, with Nanopore-based sequencing requiring as little as 1 ng [26] make this technology available to the analysis of biopsy samples with low biomass.

2 Materials

2.1 Essential Equipment

1. Class II microbiological safety cabinet.
2. Vortex.
3. Pipettes ranging from 100 to 1000 μL.
4. Balance (1.0 mg to 100 g).
5. Freezers (−80 °C and − 20 °C).
6. Microcentrifuge.
7. Appropriate personal protective equipment (*see* **Note 1**).
8. Cool box with ice packs.
9. Tablet and software for recording and curating sample collection.

2.2 Materials for Home Sample Collection Kits

1. Gloves and hand sanitizer.
2. Sample labels and metadata postcard (*see* **Note 2**).
3. IATA Packing Instruction 650-compliant packaging for transport of biologically hazardous material that is self-addressed with prepaid postage (*see* **Note 3**).
   - i. Sample tube, swab tube, etc. (primary containment).
   - ii. Resealable bag with absorbent pad (25–50 mL capacity).
   - iii. Dangerous Goods (DG) PathoSeal™ 95 A5 or A4 size bag (secondary containment).
   - iv. Return-addressed IATA-compliant shipping box such as PathoShield™ shipping box (tertiary containment).
   - v. Prepaid postage for next-day delivery.

2.3 Materials for Sample Processing

1. 70% Ethanol solution or Bioguard™
2. Disposable pipette tips.
3. Eppendorf tube, sizes from 1 to 2 mL (*see* **Note 4**).
4. Cryogenic vials, sizes from 1 to 2 mL (*see* **Note 4**).
5. Cryogenic labels.
6. Cryogenic storage boxes.

2.4 Oral Sample Home Collection and Processing Materials

1. DNA/RNA Shield SafeCollect™ Saliva Collection Kit (Zymo Research®).
2. DNA/RNA Shield Safe Collect™ Swab Kit (Zymo Research®)—1 swab kit for each site.
3. Sponge swabs Plain Toothette™, sterile, single use.
4. Sterile disposable scissors.
5. Normal saline (0.9% isotonic solution), 5-mL sterile pods.

6. Eppendorf Conical Tubes 25 mL.
7. 2 mL Lysing Matrix E™ Tubes (MP Bio®)
8. Screw top cryovials, from 1 to 2 mL (RNase, DNase, human DNA, and Pyrogen Free).
9. Cryogenic storage boxes.

2.5 Skin Sample Home Collection and Processing Materials

1. Sterilin® black top pre-wetted charcoal swabs (M40-A2, Thermo Scientific®).
2. Under the counter freezer (≥ -15 °C) (*see* **Note 5**).
3. Sterile/single-use scalpels.
4. 2 mL Lysing Matrix E™ Tubes
5. Single-use forceps.
6. Phosphate Buffer Saline (PBS), sterile, molecular grade.
7. Milli-Q® Water.
8. 40% sterile glycerol.

2.6 Vaginal Sample Collection Materials

1. FLOQSwabs™ flocked swabs with breakpoint (Copan Italia SpA).
2. 2 mL eNAT™ Guanadine Medium in 12 × 80 mm Tube (Copan Italia SpA).

2.7 Intestinal Biopsy Sample Collection Materials

1. Standard bowel preparation.
2. Biopsy forceps.
3. Sterile cryovial tubes.
4. Liquid nitrogen/dry ice.

3 Methods

3.1 Packaging Samples for Shipping

1. After sample collection, record the date and time on the label affixed to the tubes.
2. Retrieve the materials for sample shipping from the home sampling kit (*see* **Note 3**).
3. Record the date and time of sample collection on the tube label and place the sample tube in the resealable bag with the absorbent pad.
4. Place this smaller bag into the larger DG PathoSeal™ 95 (secondary containment). Seal and label the outer bag.
5. Place the sealed bag into the IATA-compliant shipping box and seal the box.
6. Affix the prepaid postage label onto the shipping box.

7. Deposit the sample box at the appropriate postal shipping facility/office for expedited delivery (*see* **Note 6**).
8. Upon receipt, prepare the Class II microbiological safety cabinet by treating the cabinet surface with Ultraviolet B (UVB) light (254 nm) for 15–30 min.
9. Turn on the cabinet laminar flow and wipe all surfaces with Bioguard™ or 70% ethanol/water solution (v/v) before and after sample processing.
10. After the samples have been processed and stored, record sample information onto REDCap® (*see* **Note 7**).

3.2 Oral Sample Collection

Some samples need to be collected by a trained professional (dentist or hygienist) during a clinic visit while others can be collected by individuals at home (*see* **Note 8**). The sample collection pathways described below were selected based upon ease of collection by study participants with no cold chain requirement.

1. Participants are provided details of activities to avoid prior to sample collection (*see* **Note 9**).
2. Saliva is collected by passive drooling and asking participants to allow saliva to accumulate at the floor of their mouth and expectorate the saliva into the saliva collection tube which should be completed within 10 min. Note the volume of saliva collected and the time needed to collect saliva.
3. Next, samples for soft tissues are collected using the DNA/RNA Shield Safe Collect Swab Kit collection tube with a Toothette™ sponge swab (*see* **Note 10**).
4. Rotate the sponge swab across the designated site.
5. Each site should be swabbed for between 15 and 60 s for regions such as the tongue where sampling might be more difficult.
6. Immediately place the swab into the sample tube by puncturing the foil seal ensuring the head of the swab is immersed in the DNA/RNA Shield solution.
7. Cut the stick of the swab using the sterile scissors and replace the screw top lid ensuring that it is secure.
8. Vigorously invert the tube containing the swab head for approximately 10 s.
9. A fresh swab should be used for each sampling site.
10. Lastly, after all sites have been collected, sample collection is completed with an oral rinse.
11. Squeeze the contents of the single-use normal saline pods into the mouth and vigorously rinse around the mouth and spit the solution into the wide-mouth 25 mL tube.

12. Using an indelible ink pen, write the date and time of the sample collection on the sample container.
13. Follow the sample mailing instructions provided.
14. Upon receiving the oral samples in the laboratory, separate the saliva and oral rinse tubes from the swabs.
15. For the saliva and oral rinse samples, vortex for 2 min at full speed.
16. Centrifuge at 14,000 *g* for 15 min and remove the supernatant.
17. Resuspend the pellet in 1.2 mL DNA/RNA shield solution and aliquot into two 400 μL cryovials and one 2 mL Lysing Matrix E™ Tube.
18. Store samples in −80 °C for later DNA extraction.
19. For the swabs, vortex for 2 min at full speed.
20. Centrifuge at 14,000 *g* for 15 min and remove the supernatant.
21. Resuspend the pellet in 1.2 mL DNA/RNA shield solution and aliquot into two 400 μL cryovials and one 2 mL Lysing Matrix E™ Tube.
22. Store samples in −80 °C for DNA extraction.
23. Record times when the participant began fasting, brushed their teeth/flossed/or used mouthwash wash and any information regarding smoking status, use of antibiotics, and probiotics.

3.3 Skin Sample Collection

The noninvasive method below is designed for skin DNA sampling from newborn infants to adults as used for the Pregnancy and EARly Life (PEARL) Study [27], and it addresses the challenge of working with low-biomass samples. The protocol includes the Promega Maxwell automated high-throughput method (Chap. 5) using magnetic beads for extracting high molecular weight DNA suitable for metagenomic-based sequencing [28].

1. Participants are provided information on activities to avoid prior to their sample collection and step-by-step instructions on how to use the collection kit (*see* **Note 11**). The collection technique can be performed at home by the participant or in a clinic by health care providers.
2. Pull and twist the charcoal swab out of its tube, taking care that the tip (swab end) does not touch anything (*see* **Note 12**).
3. Holding the swab in one hand, roll the opposite end (swab end) along the inside of the participant's forearm, ensuring that the whole of the swab end meets the skin.
4. Immediately replace the swab back into the tube, ensuring the lid is tightly closed (*see* **Note 13**).
5. Using an indelible ink pen, write the date of the sample collection on the sample container.

6. Place the sample in the freezer (−20 °C) immediately after collection, or alternatively, samples can be collected by a courier following cold chain protocols (*see* **Note 14**).
7. Upon receiving the skin sample in the laboratory, cut each swab head, using a sterile scalpel, into a 1.5-mL Eppendorf containing 1 mL of sterile PBS.
8. Vortex for 2 min at full speed.
9. Centrifuge at 14,000 *g* for 15 min and remove the supernatant.
10. Resuspend cells in 500 μL of PBS.
11. Centrifuge the remaining 300 μL at 14,000 g for 15 min and remove the supernatant:
    i. Add 100 μL of sample to each of two tubes containing 1 mL of 40% glycerol.
    ii. Store both vials as back-up aliquots of the sample at −80 °C.
12. Resuspend the pellet in 400 μL PBS.
13. Transfer to 2 mL Lysing Matrix E™ Tubes, along with the swab head using sterile forceps.
14. Store samples at −80 °C until ready for DNA extraction.

3.4 Vaginal Sample Collection

The sample collection method described is from a pilot study undertaken at the Quadram Institute [29].

1. Participants are provided instructions for home sample collection (*see* **Note 15**) and procedures to prepare for sample collection (*see* **Note 16**).
2. Place the tube filled with eNAT™– buffer on a stable, solid surface (*see* **Note 17**).
3. Open the tube cap and extract the swab from its packaging. To prevent contamination, do not touch the swab tip.
4. Spread the labia with one hand, and insert the swab 3–5 cm into the vaginal orifice using the other hand (*see* **Note 18**).
5. Rotate the swab along the vaginal wall for 10–15 s (*see* **Note 19**).
6. Remove the swab from the vagina and break the swab at the breakpoint on the tube edge.
7. Place the swab directly into the buffer, ensuring it does not contact any surfaces.
8. Close the tube with the cap, ensuring it is firmly sealed.
9. Complete the metadata postcards (*see* **Note 2**) and post with the sample (*see* **Note 20**).

10. Upon receipt in the laboratory, vortex the tube containing the swab and 2 mL eNAT™ vigorously for 60 s.
11. Aliquot 1 mL into a cryovial.
12. Store both tube and cryovial at −20 °C (*see* **Note 21**).

3.5 Colonic Biopsy Sample Collection

Colonic pinch biopsies are most used for microbiome analysis at the Quadram Institute. Colonoscopy requires specific procedures for participant preparation and specialized medical equipment used by trained professionals (*see* **Note 22**).

1. Patients are provided detailed instructions on preparing for the procedure (*see* **Note 23**).
2. During colonoscopy, the colon is examined for any growths and lesions with biopsies collected from the various macroscopically healthy areas of the colon (*see* **Note 24**).
3. Sampling time and sites are recorded for each biopsy collected.
4. Biopsies are transferred from the biopsy forceps into sterile cryovial tubes and are snap-frozen in liquid nitrogen or dry ice immediately after collection (*see* **Note 25**).
5. Samples are stored at −80 °C within 4 h of collection prior to DNA extraction.

4 Notes

1. Anyone handling human samples should undergo occupational health screening to ensure they have received the appropriate vaccinations and training. Appropriate PPE (Personal Protective Equipment) to protect the wearer from the risk of injury or infection will vary with the type of samples collected.
2. When preparing sample collection kits, provide enough sample labels for participants to record the date and time samples are collected. Use a unique pseudo-anonymized identifier to ensure anonymity of the sample during travel and during laboratory sample handling. Metadata postcards are useful for incidental information for participants to record the time when they have showered or brushed their teeth.
3. IATA (International Air Transport Association) provides guidance on the safe transport of biologically hazardous materials (UN3373). This guidance is called the Packing Instructions 650. Under this guidance, samples need three layers of containment. The primary and secondary containment must be watertight and leakproof. Absorbent material must be wrapped around the primary containment and sufficient to absorb the volume of the sample if a leak were to occur. The tertiary containment, outer packaging, serves to protect the secondary

containment during transport [30, 31]. Additionally, the package must withstand a 1.2-meter drop without leaking. When liquids are being transported, primary or secondary packaging must withstand an internal pressure of 95 kPa and temperatures between −40 and + 55 °C [30].

4. Ensure that samples are processed and stored in tubes that are RNase, DNase, human DNA, and pyrogen-free.
5. Under-counter freezers are ideal due to their small size, are minimally intrusive, and have low running costs. For the PEARL study, these freezers were used throughout the study to collect multiple samples prior to collection [27].
6. Participant instructions should specify that samples should be collected in the morning from Monday to Thursday to ensure next-day mail/courier delivery. Samples should be mailed within a maximum of four hours after collection.
7. REDCap® (Research Electronic Data Capture) is a web-based software that permits real-time data entry equipped with an audit trail. Other laboratory information management systems (LIMS) software can be used to curate the samples collected. It is best practice to reduce the number of freeze-thaw cycles to preserve sample integrity.
8. Collection of high-quality oral samples relies on clear instructions to ensure participants are well-prepared for their sample collection [32]; sample collection time needs to be consistent to minimize diurnal variation [33]; appropriate tools are used to collect enough material [5, 34, 35]; correct buffers/preservatives are used to minimize sample degradation during handling and transport [36, 37]. There is little or no consensus regarding inclusion or exclusion criteria for oral microbiome research. Likewise, there is no consensus on whether participants should abstain from brushing teeth, mouthwash, flossing, smoking, or eating/drinking prior to sample collection. Recommendations have been made to standardize patient preparation and dental examination as well as recommending using the NIH Human Microbiome Project protocol for oral sampling [32]. Participants should fast and abstain from brushing their teeth (this includes the use of mouthwash and flossing) for 12 h before sampling. Participants can drink plain water and smoke up to 1 h prior to sampling.
9. Participants are provided step-by-step instructions on how to use the collection kit to collect their oral samples. Multiple sites can be collected from the mouth, and the order in which they need to be collected is critical to prevent cross-contamination of the samples. Unless the study design requires that the sample collection order is randomized, the goal of oral sampling is to minimize cross-contamination with the sample collected

accurately reflecting the sampling site. Collection of oral samples in the following order minimizes cross-contamination: saliva, tongue, hard palate (roof of mouth), buccal mucosa (inner cheek), gingiva (gums), palatine tonsils (tonsils/throat), teeth, and oral rinse.

10. The swab material can range from cotton to synthetic materials as well as different textures of swab/brush such as flocked, sponge, and brushes. Toothette™ was selected for its large surface area and gentle abrasion from the sponge material. Alternatively, swabs included in the DNA/RNA Shield Safe-Collect™ Kit can be used. Likewise, Catch-All™ Sample Collection Swab have been used in The Human Microbiome Project [32].
11. Collection and storage instructions include collecting the samples, a timeline of when they need to be collected, how to collect them (with photographs), and information regarding sample pick up from participant homes and contact information for any questions. Participants should be instructed to avoid showering, bathing, or using any perfume or cream on their arms for 8–10 h before collecting the skin swab since they can all affect the growth of surface microorganisms and introduce fluctuations in skin microbiome populations [38].
12. Charcoal cotton swabs were selected for their suitability for multiple-site sampling, including skin, vagina, and rectum. The swab may leave a grey residue on the forearm; therefore, once sampling is complete, the sampling area should be washed with soap and water. The charcoal allows for long-term preservation of microbes at −80 °C [27]. Alternatively, OMNIgene™ SKIN (OMR-140) collection device eliminates the need for immediate freezing and allows samples to be held at ambient temperature for up to 30 days with the addition of a stabilization buffer (sodium dodecyl sulphate). However, these devices have yet to be compared with other published methods [39].
13. The use of a wet swab in a field setting or a participant's home may capture a larger number of environmental microbes, as compared to using a dry swab which is important to note when analyzing sequencing data [20]. Skin microbiome samples contain a large proportion of human (host) DNA and have low biomass. Therefore, it is vital to keep contamination as low as possible, including from the surrounding environment. A comprehensive approach is to consider sampling the participants' homes to capture their exposome/envirome [40].
14. Immediate freezing of samples at −20 °C helps stabilize microbiome communities, prevents enzymatic activity and maintains the integrity of DNA/RNA. However, this is for short-term

storage using the charcoal swabs. For long-term storage, frozen samples need to be stored at −80 °C [41].

15. This self-sampling method enables higher participation engagement, as samples can be collected at the participant's home, and forgoes the need for a gynecological examination in a clinic. Samples collected during gynecological examinations using gel lubrication can negatively impact specimen quality and microbial read count [42]. Patient-collected samples are a necessity in rural or remote areas where access to healthcare is limited. If patients have recurring genital tract infections that require frequent monitoring, this method is also advantageous in tracking changes over time and for consistent data collection. The tubes contain eNAT ™, a guanidine-thiocyanate-based media used to stabilize and preserve nucleic acids (RNA/DNA) for prolonged time periods at ambient temperatures [43]. eNat™ has been validated for microbiome analysis, and the accompanying FLOQSwabs® ensure superior elution efficiency.
16. Participants should avoid the following at least 48 h before sampling: douching, intercourse, contraceptive use, spermicides, diaphragms, cervical caps, contraceptive sponges, suppositories, feminine sprays, and genital wipes. Menstrual blood flow should have stopped at least 48 h before sampling. All participants should be verbally given clear instructions by the same researcher.
17. eNat™ reagent lyses cell membranes, completely inactivating microbes, and is therefore incompatible with methods to isolate and propagate viable microbes.
18. For positioning, the participants can be advised to assume a position for tampon insertion. Options include standing with one leg elevated on a stable surface (like a chair or toilet seat) as well as standing or sitting with knees apart.
19. Vaginal samples have relatively low biomass; collecting multiple samples from individual participants at a single time point will enhance replicability and data available for future studies.
20. Recommended metadata to be collected is the following: smoking status, alcohol consumption, contraceptive use within the preceding 12 months, last menstrual period, and any recent (within the last 3 months) antibiotic use. If participants are unable to mail their sample, they could be provided a mini freezer to store samples prior to collection.
21. Where immediate processing of samples is not possible, freezing at −20 °C or − 80 °C for up to 6 months is stated in the product brochure. Sample storage temperatures of −20 °C and − 80 °C have no significant effect on the microbial or metabolic composition of vaginal specimens [41].

22. Obtaining biopsy samples presents various challenges including the need for a skilled professional, precise protocols for processing of the samples, adhering to diet restriction and bowel preparation (laxatives) prior to colonoscopy, and low biomass of the samples [25]. With improvements in sequencing technology, the quantity of DNA required for metagenomics is now less than one nanogram [26].
23. Patients will be instructed to follow a specific regimen to empty the colon of fecal material. Two days before the colonoscopy, food intake consists of a bland, low-fiber diet. At 24 h before colonoscopy, participants continue with the restricted diet and will consume a bowel cleansing solution (e.g., sodium picosulfate (Picolax® or CitraFleet®), or if the individual has any renal insufficiency, then polyethylene glycol (PEG)-based (Klean-Prep® or Moviprep®) will be prescribed. On the day of the procedure, no liquids or food are to be taken by mouth prior to the procedure. Additionally, red-colored foods and iron supplements need to be avoided.
24. The clinician will discuss the possible risks of the procedure with the patient and offer painkillers or sedation and will obtain informed consent for providing samples for research purposes. It is also worth noting that any antibiotic use 3 months prior to the biopsies, high colonics, or probiotics will adversely influence the gastrointestinal tract (GIT) microbiome composition, and this information may not be available for the researcher or on the individual's medical records.
25. Differences in sampling methods, sample handling, and processing can impact the data collected [25, 44]. The best sampling method is non-invasive with minimal cross-contamination that provides a profile of gut microbes in different regions of the GIT. Thus, there are limitations to the use of colon biopsies for microbiome studies.

Acknowledgements

This work was funded by the Biotechnology and Biological Sciences Research Council (BBSRC) through an Institute Strategic Programme award to the QIB Programmes Gut Microbes and Health BB/R012490/1 and Food, Microbiome and Health [BB/X011054/1] and their constituent project(s).

References

1. Hou K, Wu ZX, Chen XY, Wang JQ, Zhang D, Xiao C, Zhu D, Koya JB, Wei L, Li J, Chen ZS (2022) Microbiota in health and diseases. Signal Transduct Target Ther 7(1):1–28. https://doi.org/10.1038/s41392-022-00974-4

2. Smith SB, Ravel J (2017) The vaginal microbiota, host defence and reproductive physiology. J Physiol 595(2):451–463. https://doi.org/10.1113/JP271694
3. Vuik FE, Dicksved J, Lam SY, Fuhler GM, van der Laan LJ, van de Winkel A, Konstantinov SR, Spaander MC, Peppelenbosch MP, Engstrand L, Kuipers EJ (2019) Composition of the mucosa-associated microbiota along the entire gastrointestinal tract of human individuals. United European Gastroenterol 7(7): 897–907. https://doi.org/10.1177/2050640619852255
4. Skowron K, Bauza-Kaszewska J, Kraszewska Z, Wiktorczyk-Kapischke N, Grudlewska-Buda K, Kwiecińska-Piróg J, Wałecka-Zacharska E, Radtke L, Gospodarek-Komkowska E (2021) Human skin microbiome: impact of intrinsic and extrinsic factors on skin microbiota. Microorganisms 9(3):543. https://doi.org/10.3390/microorganisms9030543
5. Baker JL, Mark Welch JL, Kauffman KM, McLean JS, He X (2024) The oral microbiome: diversity, biogeography and human health. Nat Rev Microbiol 22(2):89–104. https://doi.org/10.1038/s41579-023-00963-6
6. Lyra A, Forssten S, Rolny P, Wettergren Y, Lahtinen SJ, Salli K, Cedgård L, Odin E, Gustavsson B, Ouwehand AC (2012) Comparison of bacterial quantities in left and right colon biopsies and faeces. World J Gastroenterol 18(32):4404. https://doi.org/10.3748/wjg.v18.i32.4404
7. Jašarević E, Howard CD, Misic AM, Beiting DP, Bale TL (2017) Stress during pregnancy alters temporal and spatial dynamics of the maternal and offspring microbiome in a sex-specific manner. Sci Rep 7(1):44182. https://doi.org/10.1038/srep44182
8. Proctor DM, Fukuyama JA, Loomer PM, Armitage GC, Lee SA, Davis NM, Ryder MI, Holmes SP, Relman DA (2018) A spatial gradient of bacterial diversity in the human oral cavity shaped by salivary flow. Nat Commun 9(1):681. https://doi.org/10.1038/s41467-018-02900-1
9. Aagaard K, Petrosino J, Keitel W, Watson M, Katancik J, Garcia N, Patel S, Cutting M, Madden T, Hamilton H, Harris E (2013) The human microbiome project strategy for comprehensive sampling of the human microbiome and why it matters. FASEB J 27(3):1012. https://doi.org/10.1096/fj.12-220806
10. Qian XB, Chen T, Xu YP, Chen L, Sun FX, Lu MP, Liu YX (2020) A guide to human microbiome research: study design, sample collection, and bioinformatics analysis. Chin Med J 133(15):1844–1855. https://doi.org/10.1097/CM9.0000000000000871
11. Mirzayi C, Renson A, Genomic Standards Consortium, Massive Analysis and Quality Control Society Furlanello Cesare 31 Sansone Susanna-Assunta 84, Zohra F, Elsafoury S, Geistlinger L, Kasselman LJ, Eckenrode K, van de Wijgert J, Loughman A (2021) Reporting guidelines for human microbiome research: the STORMS checklist. Nat Med 27(11): 1885–1892. https://doi.org/10.1038/s41591-021-01552-x
12. Welch JL, Ramírez-Puebla ST, Borisy GG (2020) Oral microbiome geography: micron-scale habitat and niche. Cell Host Microbe 28(2):160–168. https://doi.org/10.1016/j.chom.2020.07.009
13. Bang E, Oh S, Ju U, Chang HE, Hong JS, Baek HJ, Kim KS, Lee HJ, Park KU (2023) Factors influencing oral microbiome analysis: from saliva sampling methods to next-generation sequencing platforms. Sci Rep 13(1):10086. https://doi.org/10.1038/s41598-023-37246-2
14. Chen H, Zhao Q, Zhong Q, Duan C, Krutmann J, Wang J, Xia J (2022) Skin microbiome, metabolome and skin phenome, from the perspectives of skin as an ecosystem. Phenomics 2(6):363–382. https://doi.org/10.1007/s43657-022-00073-y
15. Salem I, Ramser A, Isham N, Ghannoum MA (2018) The gut microbiome as a major regulator of the gut-skin axis. Front Microbiol 9: 1459. https://doi.org/10.3389/fmicb.2018.01459
16. Constantinou A, Kanti V, Polak-Witka K, Blume-Peytavi U, Spyrou GM, Vogt A (2021) The potential relevance of the microbiome to hair physiology and regeneration: the emerging role of metagenomics. Biomedicine 9(3):236. https://doi.org/10.3390/biomedicines9030236
17. Ying S, Zeng DN, Chi L, Tan Y, Galzote C, Cardona C, Lax S, Gilbert J, Quan ZX (2015) The influence of age and gender on skin-associated microbial communities in urban and rural human populations. PLoS One 10(10):e0141842. https://doi.org/10.1371/journal.pone.0141842
18. Ozkan J, Willcox M, Coroneo M (2022) A comparative analysis of the cephalic microbiome: the ocular, aural, nasal/nasopharyngeal, oral and facial dermal niches. Exp Eye Res 220:109130. https://doi.org/10.1016/j.exer.2022.109130
19. Prast-Nielsen S, Tobin AM, Adamzik K, Powles A, Hugerth LW, Sweeney C, Kirby B, Engstrand L, Fry L (2019) Investigation of the

skin microbiome: swabs vs. biopsies. Br J Dermatol 181(3):572–579. https://doi.org/10.1111/bjd.17691

20. Manus MB, Kuthyar S, Perroni-Marañón AG, de la Mora AN, Amato KR (2022) Comparing different sample collection and storage methods for field-based skin microbiome research. Am J Hum Biol 34(1):e23584. https://doi.org/10.1002/ajhb.23584
21. Soga N, Nagura R, Nakamura R, Kohda K, Motoyama Y, Ito M, Nakagawa I, Nakajima S, Ikeuchi A (2024) Simple tape-stripping method for highly reliable and quantitative analysis of skin microbiome. Exp Dermatol 33(8):e15154. https://doi.org/10.1111/exd.15154
22. Brooks JP, Edwards DJ, Harwich MD, Rivera MC, Fettweis JM, Serrano MG, Reris RA, Sheth NU, Huang B, Girerd P, Vaginal Microbiome Consortium (2015) The truth about metagenomics: quantifying and counteracting bias in 16S rRNA studies. BMC Microbiol 15:66. https://doi.org/10.1186/s12866-015-0351-6
23. Sharaf RN, Shergill AK, Odze RD, Krinsky ML, Fukami N, Jain R, Appalaneni V, Anderson MA, Ben-Menachem T, Chandrasekhara V, Chathadi K (2013) Endoscopic mucosal tissue sampling. Gastrointest Endosc 78(2):216–224. https://doi.org/10.1016/j.gie.2013.04.167
24. Virine B, Chande N, Driman DK (2020) Biopsies from ascending and descending colon are sufficient for diagnosis of microscopic colitis. Clin Gastroenterol Hepatol 18(9):2003–2009. https://doi.org/10.1016/j.cgh.2020.02.036
25. Tang Q, Jin G, Wang G, Liu T, Liu X, Wang B, Cao H (2020) Current sampling methods for gut microbiota: a call for more precise devices. Front Cell Infect Microbiol 10:151. https://doi.org/10.3389/fcimb.2020.00151
26. Simon SA, Schmidt K, Griesdorn L, Soares AR, Bornemann TL, Probst AJ (2023) Dancing the Nanopore limbo–Nanopore metagenomics from small DNA quantities for bacterial genome reconstruction. BMC Genomics 24(1):727. https://doi.org/10.1186/s12864-023-09853-w
27. Phillips S, Watt R, Atkinson T, Savva GM, Hayhoe A, Hall LJ, PEARL Study Team (2021) The pregnancy and EARly life study (PEARL)-a longitudinal study to understand how gut microbes contribute to maintaining health during pregnancy and early life. BMC Pediatr 21(1):357. https://doi.org/10.1186/s12887-021-02835-5
28. Serghiou IR, Baker D, Evans R, Dalby MJ, Kiu R, Trampari E, Phillips S, Watt R, Atkinson T, Murphy B, Hall LJ (2023) An efficient method for high molecular weight bacterial DNA extraction suitable for shotgun metagenomics from skin swabs. Microb Genom 9(7):001058. https://doi.org/10.1099/mgen.0.001058
29. Gao XS, Groot T, Schoenmakers S, Louwers Y, Budding A, Laven J (2024) The vaginal microbiome: patient- versus physician-collected microbial swab: a pilot study. Microorganisms 12(9):1859. https://doi.org/10.3390/microorganisms12091859
30. IATA (2025) Title of subordinate document. In: IATA dangerous goods regulations, 66th edn. International Air Transport Association. https://www.iata.org/contentassets/b08040a138dc4442a4f066e6fb99fe2a/dgr-66-en-pi650.pdf. Accessed 10 Apr 2025
31. World Health Organization (2021) Section 6: preparing packaging requirements. In: Guidance on regulations for the transport of infectious substances 2021–2022. World Health Organization. https://cdn.who.int/media/docs/default-source/influenza/global-influenza-surveillance-and-response-system/who_guidance_is_2021-22.pdf?sfvrsn=5f67b03e_3. Accessed 10 Apr 2025
32. Brzychczy-Sroka B, Talaga-Ćwiertnia K, Sroka-Oleksiak A, Gurgul A, Zarzecka-Francica E, Ostrowski W, Kąkol J, Drożdż K, Brzychczy-Włoch M, Zarzecka J (2024) Standardization of the protocol for oral cavity examination and collecting of the biological samples for microbiome research using the next-generation sequencing (NGS): own experience with the COVID-19 patients. Sci Rep 14(1):3717. https://doi.org/10.1038/s41598-024-53992-3
33. Takayasu L, Suda W, Takanashi K, Iioka E, Kurokawa R, Shindo C, Hattori Y, Yamashita N, Nishijima S, Oshima K, Hattori M (2017) Circadian oscillations of microbial and functional composition in the human salivary microbiome. DNA Res 24(3):261–270. https://doi.org/10.1093/dnares/dsx001
34. Rogers NL, Cole SA, Lan HC, Crossa A, Demerath EW (2007) New saliva DNA collection method compared to buccal cell collection techniques for epidemiological studies. Am J Hum Biol 19(3):319–326. https://doi.org/10.1002/ajhb.20586
35. Omori M, Kato-Kogoe N, Sakaguchi S, Fukui N, Yamamoto K, Nakajima Y, Inoue K, Nakano H, Motooka D, Nakano T, Nakamura S (2021) Comparative evaluation of microbial profiles of oral samples obtained at different collection time points and using different

methods. Clin Oral Investig 25:2779–2789. https://doi.org/10.1007/s00784-020-03592-y

36. Zhou X, Nanayakkara S, Gao JL, Nguyen KA, Adler CJ (2019) Storage media and not extraction method has the biggest impact on recovery of bacteria from the oral microbiome. Sci Rep 9(1):14968. https://doi.org/10.1038/s41598-019-51448-7
37. Vogtmann E, Chen J, Kibriya MG, Amir A, Shi J, Chen Y, Islam T, Eunes M, Ahmed A, Naher J, Rahman A (2019) Comparison of oral collection methods for studies of microbiota. Cancer Epidemiol Biomarkers Prev 28(1): 137–143. https://doi.org/10.1158/1055-9965.EPI-18-0312
38. Kong HH, Andersson B, Clavel T, Common JE, Jackson SA, Olson ND, Segre JA, Traidl-Hoffmann C (2017) Performing skin microbiome research: a method to the madness. J Invest Dermatol 137(3):561–568. https://doi.org/10.1016/j.jid.2016.10.033
39. Bouevitch A, Macklaim J, Le François B (2020) OMNIgene® SKIN (OMR-140): an optimized collection device for the capture and stabilization of the human skin microbiome. DNA 20:200
40. Khmaladze I, Leonardi M, Fabre S, Messaraa C, Mavon A (2020) The skin interactome: a holistic "genome-microbiome-exposome" approach to understand and modulate skin health and aging. Clin Cosmet Investig Dermatol 24:1021–1040. https://doi.org/10.2147/CCID.S239367
41. Hubel A, Spindler R, Skubitz AP (2014) Storage of human biospecimens: selection of the optimal storage temperature. Biopreserv Biobank 12(3):165–175. https://doi.org/10.1089/bio.2013.0084
42. Amitai KD, Hadar R, Paulson JN, Mordechai Y, Eskandarian HA, Efroni G, Amir A, Haberman Y, Tsur A (2024) Lubricating gel influence on vaginal microbiome sampling. Sci Rep 14(1):18223. https://doi.org/10.1038/s41598-024-68948-w
43. Bai G, Gajer P, Nandy M, Ma B, Yang H, Sakamoto J, Blanchard MH, Ravel J, Brotman RM (2012) Comparison of storage conditions for human vaginal microbiome studies. PLoS One 7(5):e36934. https://doi.org/10.1371/journal.pone.0036934
44. Nowicki C, Ray L, Engen P, Madrigrano A, Witt T, Lad T, Cobleigh M, Mutlu EA (2023) Comparison of gut microbiome composition in colonic biopsies, endoscopically-collected and at-home-collected stool samples. Front Microbiol 14:1148097. https://doi.org/10.3389/fmicb.2023.1148097

Chapter 5

DNA Extraction and Quality Control

Rachel Watt, Sarah Phillips, Jennifer Ahn-Jarvis, Gemma Beasy, and Iliana Serghiou

Abstract

Understanding the role of the human microbiome in health and disease is dependent upon acquiring high-quality DNA from host samples including feces, tissue biopsies, and mucosal swabs. DNA extraction methods need to be effective for downstream applications including metagenomic sequencing analysis. This chapter describes the best practice for DNA extraction for microbiome analysis of human fecal, oral, low vaginal, skin swabs, and colon biopsy samples, and quality control assessment using Nanodrop and Qubit technology.

Key words DNA extraction, DNA purification, Quality control, Nanodrop, Qubit

1 Introduction

The complex human gut microbiome contains microbes that cannot be maintained in vitro with research instead relying on obtaining DNA of a high enough quality and in sufficient amounts for phylogenetic profiling by genome sequencing [1].

The factors to consider when choosing DNA extraction methods include downstream processing methodology and the sample source. Effective DNA extraction is reliant on optimized sample preparation, specifically, the successful lysing of all organisms particularly within fecal samples [1]. Gram-negative organisms more readily lyse with lysozyme and chelating agents (e.g., ethylenediaminetetraacetic acid, EDTA). However, Gram-positive bacteria and fungi are more resistant to lysis [2]. Also, fecal samples contain food debris that can impact lysis. To overcome these issues, a combination of chemical and mechanical lysis (bead beating) can be used with fecal samples to ensure comprehensive microbial lysis. Further enhancements include the use of high-speed bead-beating devices (e.g., MpBio Fastprep) in combination with Lysing E matrix tubes to ensure lysis of more resistant microorganisms and decreased processing time [3]. High-throughput, automated methods of

Simon R. Carding (ed.), *Best Practice in Microbiome Research*, Springer Protocols Handbooks, https://doi.org/10.1007/978-1-0716-5009-7_5, © The Author(s) 2026

DNA extraction using robotic-based systems such as Promega Maxwell® are ideal for sample processing in large-scale microbiome studies.

Isolation of high-quality bacterial DNA from buccal, skin, and vaginal samples presents a unique set of challenges which need to be considered to ensure that results are reproducible and robust. Samples from these locations are susceptible to sampling variability (location and collection tools) [4–6] and environmental contamination [5]. Moreover, these regions typically have low microbial biomass [7, 8] with high host (human) DNA contamination [8–10] and bacteria resistant to cell lysis [11] together with the presence of endogenous DNA-degrading enzymes or PCR-inhibitory factors [12–14].

For skin, oral, and vaginal samples, mechanical, chemical, and/or enzymatic lysing agents are preferred. Bead beating alone or with chemical lysis and alkaline solutions (e.g., sodium hydroxide) or detergents (e.g., sodium dodecyl sulphate) as well as enzymes such as lysozyme or proteinases are effective in lysing bacteria [15].

For colonic biopsies, the DNA extraction method should include methods to deplete host DNA [16]. At the Quadram Institute, we have adopted a protocol that uses the QIAamp® PowerFecal Pro DNA Kit for DNA extraction from biopsy tissues [17] with modifications as summarized in Fig. 1.

2 Materials

2.1 Essential Equipment and Reagents

2.1.1 DNA Extraction from Fecal Samples Using Maxwell® RSC PureFood GMO and Authentication Kit

1. MpBio Lysing Matrix E tube.
2. MpBio Fast prep.
3. Promega Maxwell® robot.
4. 1 ml Gilson pipette
5. Vortex.
6. Sterile, aerosol-resistant pipette tips.
7. Heat block.
8. Microcentrifuge.

2.1.2 Maxwell® RSC PureFood GMO and Authentication Kit Reagents

1. RNase A solution.
2. Elution Buffer.
3. Maxwell® RSC cartridge.
4. Elution tubes (0.5ml).
5. CTAB Buffer.
6. Proteinase K solution.

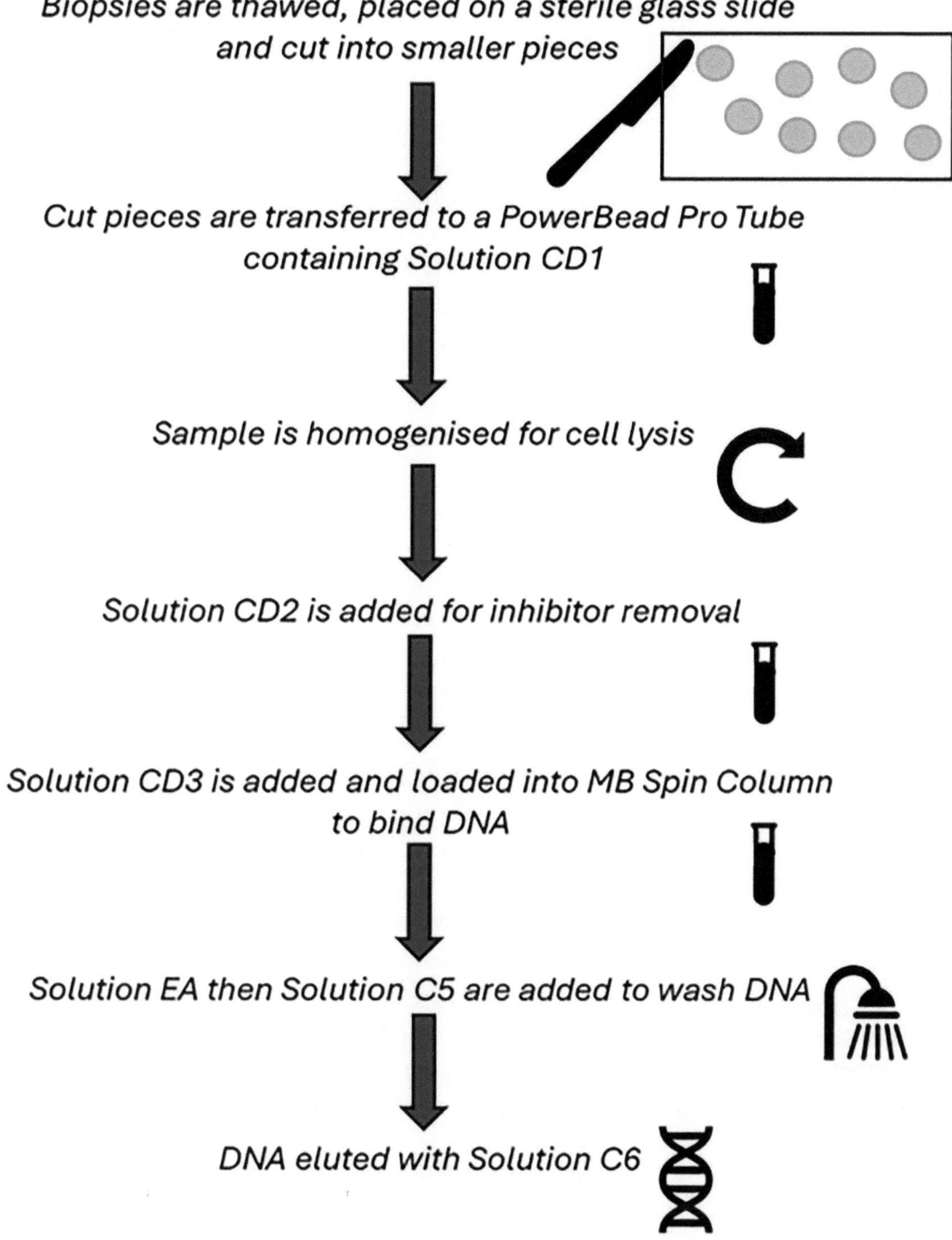

Fig. 1 DNA extraction of colonic biopsies using QIAamp® PowerFecal Pro DNA Kit

7. Lysis buffer.
8. RSC plungers.

2.2 DNA Extraction of Colon Biopsies Using QIAamp® PowerFecal Pro DNA Kit (Fig. 1)

Essential Equipment

1. 2 mL microcentrifuge tubes/2 mL collection tubes.
2. 1.5 mL elution tubes.
3. PowerLyzer 24 Homogenizer (Qiagen: 13155).
4. Microcentrifuge (up to 16000× *g*, at room temperature 20 °C, Eppendorf®: 5425R).
5. Vortex Genie® 2 (Z258423).
6. Pipettes (from 50 to 1000 μL, Eppendorf®: 3123000942).
7. MSC class II cabinet.
8. Frozen biopsies.
9. Sterile glass slides.
10. Sterile scalpel.

2.2.1 QIAamp® PowerFecal Pro DNA Kit (Qiagen: 51804, 50 Samples)

Reagents

1. Quick Start Protocol.
2. 50 PowerBead Pro Tubes.
3. 50 MB Spin Columns.
4. 40 mL Solution CD1.
5. 15 mL Solution CD2.
6. 35 mL Solution CD3.
7. 36 mL Solution EA.
8. 30 mL Solution C5.
9. 9 mL Solution C6.
10. Disinfectants: 70% ethanol in water, Bioguard, Rely On Virkon tablets.

2.3 DNA Extraction from Oral Samples, Low Vaginal, and Skin Swabs Using Promega Maxwell® Robot

Essential Equipment

1. Promega Maxwell® robot.
2. Class II microbiological safety cabinet.
3. Vortex (Genie 2 Vortexer; Scientific Industries Bohemia, USA).
4. Micro Centrifuge (Eppendorf, Hamburg, GER).
5. ThermoMixer® C (Eppendorf, Hamburg, GER).
6. Tissue Lyser (Qiagen, Venlo, NLD).

Reagents

1. 2 ml Lysing Matrix E Tubes (MP Biomedicals, Santa Ana, USA).

2. Maxwell® RSC Blood DNA Kit (Promega, Madison, USA).
3. Thermo Fisher Lysozyme working solution.
4. Tris EDTA Buffer Solution 7.4 pH, Catalogue No. 93302 (BioUltra, Merck, Rahway, USA).
5. Qiagen Buffer ATL, Catalogue No. 19076, (Qiagen, Venlo, NLD).
6. Qiagen Buffer ACL, Catalogue No. 939017 (Qiagen, Venlo, NLD).
7. QIAmp Viral RNA mini–Accessory Set (Qiagen, Venlo, NLD).
1. Carrier RNA.
2. AVE Buffer.

2.4 DNA Quantification and Quality Control

2.4.1 Invitrogen Qubit Fluorometer

Essential Equipment

1. Invitrogen Qubit Fluorometer.
2. Qubit assay tubes.
3. Gilson Pipette.
4. MSC class II cabinet.

Reagents

1. dsDNA BR Reagent 200X concentrate in DMSO.
2. dsDNA BR Buffer,
3. dsDNA BR Standard #1, 0 ng/μL,
4. dsDNA BR Standard #2, 100 ng/μL,
5. dsDNA HS Reagent 200X concentrate in DMSO.
6. dsDNA HS Buffer,
7. dsDNA HS Standard #, 0 ng/μL,
8. dsDNA HS Standard #2, 100 ng/μL.

2.4.2 Nanodrop Spectrophotometer

Essential Equipment

1. Nanodrop and a computer with installed software.
2. 10 μL pipette
3. Pipette tips.
4. Lint-free lab wipes.

Reagents

1. EB buffer (used as a blank). This should be whichever liquid the DNA is resuspended in.
2. Distilled water.

3 Methods

3.1 Fecal DNA Extraction Using the Promega Maxwell® Robot

(*See* **Note 1** for Additional Methodological Details)

1. Place 200 mg of feces into a lysing matrix E tube with a screw top lid.
2. Add 1 ml CTAB to the sample and vortex for 30 s.
3. Heat the sample in a dry heating block at 95 °C for 5 min.
4. Vortex for 1 min.
5. Homogenize in the "FastPrep" Instrument for 45 s at 6.0 m/s.
6. Pulse centrifuge the samples.
7. Add 40 μl of proteinase K and 20 μl of RNase A to the samples and vortex to mix.
8. Heat the sample in a dry heat block at 70 °C for 10 min.
9. Centrifuge the sample at 14000× *g* for 5 min to pellet debris.
10. Set up the Maxwell® instrument (*see* **Note 1**).

3.2 Colonic Biopsy DNA Extraction Using QIAamp® PowerFecal Pro DNA Kit

See Fig. 1 and **Note 2** for an overview and additional methodological details.

1. Remove the frozen biopsies from the −80 °C freezer and thaw.
2. Place the biopsies on a sterile glass slide.
3. Using a sterile scalpel, cut the biopsies into smaller pieces.
4. Transfer to a PowerBead Pro Tube containing 800 μL of Solution CD1. The PowerBead Pro Tube can be briefly centrifuged if the beads are not at the bottom.
5. Vortex the PowerBead Pro Tubes (containing the tissues and CD1 solution) briefly and place in the homogenizer.
6. Homogenize the sample by bead beating at 2000 rpm for 30 s, pausing for 30 s, then homogenize again at 2000 rpm for 30 s.
7. Centrifuge at 15000× *g* for 1 min.
8. Transfer supernatant into a clean 2 mL microcentrifuge tube. *Proceed to inhibitor removal.*
9. Add 200 μL of Solution CD2 to the supernatant and vortex briefly.
10. Centrifuge at 15000× *g* for 1 min.
11. Transfer supernatant (~500–600 μL) into a clean 2 mL microcentrifuge tube. *Proceed to DNA binding.*
12. Add 600 μL of Solution CD3 to the supernatant and vortex briefly. Solution CD3 can be heated at 60 °C to remove precipitation (if this has occurred).

13. Transfer 650 μL of the lysate and place it into an MB Spin Column.
14. Centrifuge at 15000× *g* for 1 min.
15. Discard the flow-through.
16. Place the remaining lysate (~450–550 μL) into an MB Spin Column.
17. Centrifuge once more at 15000× *g* for 1 min.
18. Transfer the MB Spin Column into a clean 2 mL microcentrifuge tube. *Proceed to the DNA wash.*
19. Add 500 μL of Solution EA to the MB Spin Column.
20. Centrifuge at 15000× *g* for 1 min.
21. Discard the flow-through.
22. Carefully place the MB Spin Column back into the same 2 mL microcentrifuge tube.
23. Add 500 μL of Solution C5 to the MB Spin Column.
24. Centrifuge at 15000× *g* for 1 min.
25. Discard the flow-through.
26. Place the MB Spin Column back into a new clean 2 mL microcentrifuge tube.
27. Centrifuge at 16000× *g* for 2 min. *Proceed to DNA elution and QC analysis.*
28. Carefully place the MB Spin Column into a new 1.5 mL elution tube.
29. Elute DNA by adding 50 μL of Solution C6 directly to the center of the white filter membrane.
30. Centrifuge at 15000× *g* for 1 min.
31. Discard the MB Spin Column.
32. The DNA will be eluted in the 1.5 mL elution tube.
33. DNA purity and concentration are evaluated using the Nanodrop spectrophotometer or Qubit.
34. Store the DNA at −20 °C until sequencing analysis.

3.3 DNA Extraction from Oral, Vaginal, and Skin Samples Using Promega Maxwell® Robot

(*See* **Note 3** for additional methodological details)

1. Using a sterile scalpel, cut each swab head into a 1.5 mL Eppendorf and add 1 mL of 1X PBS.
2. Vortex for 2 min on full speed.
3. Centrifuge at 14000× *g* for 15 min and remove supernatant.
4. Resuspend cells in 400 μL of 1 X PBS and transfer to a 2 mL Lysing Matrix E Tube along with the swab head.

5. Add 3 μL of lysozyme solution at 250 U/μL. A 300 μL lysozyme working solution is prepared by adding 75 μL of lysozyme to 225 μL of Tris EDTA buffer.
6. Incubate with agitation at 300 rpm at 37 °C for 20–22 h in a dry heat block.
7. After incubation, bead beat using a Tissue Lyser for 3 min at 20 Hz.
8. Add the following to the lysing tube:
   (i) 40 μL proteinase K.
   (ii) 165 μL buffer ATL.
   (iii) 120 μL carrier RNA—first, reconstitute the lyophilized carrier RNA using buffer AVE to create a 1 μg/μL solution.
   (iv) 315 μL buffer ACL.
9. Mix and incubate at 68 °C for 15 min.
10. Centrifuge at 14000× *g* for 15 min.
11. Maxwell® instrument set-up (*see* **Note 1**). Select a method by scanning or entering the 2D bar code on the kit box to select the "RSC Blood DNA" and "Authentication method" and touch "proceed."
12. Store the extracted DNA short-term for up to 2 weeks at 2–10 °C and long-term for up to 2 years at −20 °C or −80 °C. For longer periods, liquid nitrogen storage should be used.

3.4 DNA Quantification and QC

(*See* **Note 4** for additional methodological details).

3.4.1 Invitrogen Qubit Fluorometer

1. Prepare the assay tubes required for samples and two additional tubes for the standards.
2. Make the working solution by diluting the qubit reagent in qubit buffer 1 in 200 μL.
3. Prepare the assay tubes as below:

	Tubes for standards	Tubes for samples
Volume of working solution	190 μL	180–199 μL
Volume of sample	N/A	1–20 μL
Volume of standard	10 μL	N/A
Total volume in each tube	200 μL	200 μL

4. Vortex for 3 s.

5. Incubate tubes at 20 °C for 2 min protecting them from light.
6. After incubation, the tubes are read on the Qubit fluorometer.
7. First, measure the standards. Follow "on screen" instructions.
8. Choose the assay appropriate for the samples using the touch screen.
9. Choose the appropriate kit type, either broad range (BR) or high sensitivity (HS).
10. Select "Read standards." When prompted, place Standard #1 into the sample chamber and press "Read standard." Remove Standard #1 and place Standard #2. Press "Read standard" and a standard curve will be automatically generated.
11. Select "Run samples" (below the standard curve). Adjust the sample volume according to how much sample was initially added to the assay tube and choose the output sample units (often ng/μL).
12. Place the sample into the sample chamber and select "Read tube." A results screen will show the calculated amount of DNA/RNA/protein present in the original sample.
13. Record or export this value.

3.4.2 Nanodrop Spectrometer

1. First, clean the instrument by adding 3 μL of purified water to the lower pedestal and then place the arm back down.
2. Lift the arm again after ~20 s and then wipe using a lint-free lab wipe. Proceed to using the instrument.
3. Select the "Nucleic Acid" button on the Nanodrop software.
4. Add 2 μL of purified water to the lower pedestal and lower the arm. Click "ok." This will initialize the nanodrop.
5. After initialization, wipe the pedestal dry with a lint-free wipe.
6. Add 2 μL of the EB buffer to create a blank for the measurements.
7. Lower the arm and click "Blank."
8. When complete, wipe the pedestal dry.
9. Add 2 μL of the sample and lower the arm and click "Measure."
10. Collect the data on interest from the screen produced.
11. Clean the nanodrop (**steps 1–2**) and repeat for all samples.
12. When all samples have been measured, clean the nanodrop and shut down the computer.

4 Notes

1. Maxwell® Instrument setup.
   - (i) Turn on the instrument and tablet PC. Log in to the Tablet PC and start the Maxwell® software by double-touching the icon on the desktop.
   - (ii) Select a method by scanning or entering the 2D bar code on the kit box to select the "PureFood GMO" and "Authentication method" and touch "proceed."
   - (iii) On the "Cartridge Setup" screen, select the rack positions of your samples. Enter any required sample numbers, and touch "proceed."
   - (iv) Place cartridges in the rack and remove foil seals. The cartridges have eight wells that are numbered. The cartridges should be placed in the rack so that well 1 is furthest away from the elution tubes.
   - (v) Add 100 μL elution buffer into each of the supplied elution tubes and place in the numbered holes of the rack. Ensure the elution tubes are labelled with the relevant sample code. It is important to only use the elution tubes that are provided with the kit as other elution tubes may not be compatible.
   - (vi) Place the plunger into the eighth well of the cartridge.
   - (vii) Add 300 μL of lysis buffer (RED label) into well 1 of the cartridges.
   - (viii) Add 300 μL of supernatant to well 1 of the cartridge and mix by pipetting up and down.
   - (ix) Place the rack with the inserted cartridges onto the Maxwell®. Ensure that the rack is placed in the Maxwell® with the elution tubes closest to the door.
   - (x) Touch "Start." The machine will take 35 min for each run.
   - (xi) Once the extraction is completed, store the DNA short-term (up to 2 weeks) at 2–10 °C and − 20 °C or − 80 °C for longer-term storage (up to 2 years) to prevent degradation of the sample. For much longer studies, DNA can be stored in liquid nitrogen to prevent degradation over decades.
2. The Promega Maxwell® uses magnetic beads that bind to particles in the first well and then carry these particles through the remaining wells during the DNA extraction process. However, this can result in high carryover of beads in the eluted DNA sample, which can affect the accuracy of DNA quantification [18]. If this does become an issue, magnetic tube stands

can be used to attract the beads while the eluted DNA is transferred to new tubes.

It is good practice to use a negative control for each DNA run. This will help identify any contaminants during sample processing with the kit described. A positive control sample using a mock community of typical gut microorganisms at known concentrations can also be included. Alternative methods for DNA extraction from fecal samples include the MPBio Fast Spin Soil Kit. However, this is a manual kit and will take longer to process large numbers of samples and will not have the same consistency as an automated method. This method may not therefore be ideal for larger microbiome studies but can produce high concentrations of DNA from small numbers of samples. If calibration fails on the Qubit, it will show a calibration error message. Troubleshoot this by rereading the standards, ensuring the correct assay has been selected. If this fails, try new sets of standards and read again.

3. The QIAamp® PowerFecal Pro DNA Kit is for the isolation of microbial DNA from gut and stool samples. DNA is extracted according to the manufacturer's protocol, and this protocol along with the solutions is included within the purchased kit (Kit Handbook/Quick-Start). DNA extraction procedures should be performed in MSC class II containment conditions using sterile equipment. The PowerBead Pro Tube contains a buffer that will begin to dissolve the inhibitors and protect nucleic acids from degradation. If the beads are not at the bottom of the PowerBead Pro Tube, they can be centrifuged and can also be vortexed to disperse the buffer and beads (**step 4**, Subheading 3.2). Homogenizing samples at higher speeds such as up to 4000 rpm may increase DNA yields but will lead to more DNA fragmentation (**steps 5** and **6**). Supernatants may contain small fragments of the biopsies; this is normal. Solution CD2 contains inhibitor removal technology, able to precipitate non-DNA materials including proteins and cell debris. It is important to remove these materials as this may affect DNA purity and quality (**step 9**). Once centrifuged, do not disturb the pellet as this contains the non-DNA materials including proteins and cell debris that you do not want to transfer (**steps 10** and **11**). Solution CD3 contains a high salt concentration so the DNA binds to the silica of the MB spin column filter membrane. This allows other materials to pass through the filter membrane while the DNA is bound to the membrane (**steps 12** and **13**). The wash buffer (Solution EA) removes nonaqueous materials and proteins from the MB spin column filter membrane (**step 19**), while the wash buffer (Solution C5) is ethanol-based to further clean the bound DNA (**step 23**). These are both needed to effectively wash

the DNA. This centrifugation step at 16000× *g*g is key to remove all solution C5 that could interfere with downstream applications such as PCR (**step 27**). It is vital to place the 50 μL of solution C6 in the center of the membrane to make sure the entire membrane is covered and wet. This will allow a better release of DNA from the filter membrane (**step 29**). As solution C6 does not contain EDTA, the DNA can be stored between −30 and − 15 °C or − 90 °C and − 65 °C; we routinely use −20 °C until sequencing analysis (**step 34**).

4. The "Qubit Fluorometer" accurately measures the quantity of extracted DNA, RNA, and protein. It uses a fluorescent dye to emit a signal when bound to the extracted DNA. This ensures that only the double-stranded DNA is quantified and that contaminants are not included in the final quantification. It is a relatively fast process conducted using a kit of pre-made reagents. There are high-sensitivity reagents (HS) as well as broad-range reagents (BR). The HS reagents should be used on samples that you assume have low quantities of DNA, e.g., DNA extracted from skin swabs. The BR reagents are not as sensitive as the HS reagents but will quantify samples with both high and low DNA concentrations. DNA extraction from fecal samples usually contains a high concentration of DNA, so BR reagents are most suitable.

Acknowledgements

This work was funded by the Biotechnology and Biological Sciences Research Council (BBSRC) through an Institute Strategic Programme award to the QIB Programmes Gut Microbes and Health BB/R012490/1 and Food, Microbiome and Health [BB/X011054/1] and their constituent project(s).

References

1. Lim MY et al (2018) Comparison of DNA extraction methods for human gut microbial community profiling. Syst Appl Microbiol 41(2):151–157. https://doi.org/10.1016/j.syapm.2017.11.008
2. d'Enfert C et al (2021) The impact of the fungus-host-microbiota interplay upon Candida albicans infections: current knowledge and new perspectives. FEMS Microbiol Rev 45(3). https://doi.org/10.1093/femsre/fuaa060
3. Whyte JD (2016) In: Micic M (ed) Bead-beating sample preparation for nucleic acids isolation from fecal samples, in sample preparation techniques for soil, plant, and animal samples. Springer, New York, pp 353–364. https://doi.org/10.1007/978-1-4939-3185-9_25
4. Hall MW et al (2017) Inter-personal diversity and temporal dynamics of dental, tongue, and salivary microbiota in the healthy oral cavity. NPJ Biofilms Microbiomes 3:2. https://doi.org/10.1038/s41522-016-0011-0
5. Bjerre RD et al (2019) Effects of sampling strategy and DNA extraction on human skin microbiome investigations. Sci Rep 9(1): 17287. https://doi.org/10.1038/s41598-019-53599-z
6. Kim TK et al (2009) Heterogeneity of vaginal microbial communities within individuals. J

Clin Microbiol 47(4):1181–1189. https://doi.org/10.1128/JCM.00854-08

7. Rosenbaum J et al (2019) Evaluation of Oral cavity DNA extraction methods on bacterial and fungal microbiota. Sci Rep 9(1):1531. https://doi.org/10.1038/s41598-018-38049-6
8. Serghiou IR et al (2023) An efficient method for high molecular weight bacterial DNA extraction suitable for shotgun metagenomics from skin swabs. Microb Genom 9(7). https://doi.org/10.1099/mgen.0.001058
9. Ahannach S et al (2021) Microbial enrichment and storage for metagenomics of vaginal, skin, and saliva samples. iScience 24(11):103306. https://doi.org/10.1016/j.isci.2021.103306
10. Marquet M et al (2022) Evaluation of microbiome enrichment and host DNA depletion in human vaginal samples using Oxford Nanopore's adaptive sequencing. Sci Rep 12(1):4000. https://doi.org/10.1038/s41598-022-08003-8
11. Gill C et al (2016) Evaluation of lysis methods for the extraction of bacterial DNA for analysis of the vaginal microbiota. PLoS One 11(9):e0163148. https://doi.org/10.1371/journal.pone.0163148
12. Shvartsman E et al (2022) Comparative analysis of DNA extraction and PCR product purification methods for cervicovaginal microbiome analysis using cpn60 microbial profiling. PLoS One 17(1):e0262355. https://doi.org/10.1371/journal.pone.0262355
13. Zhou X et al (2019) Storage media and not extraction method has the biggest impact on recovery of bacteria from the oral microbiome. Sci Rep 9(1):14968. https://doi.org/10.1038/s41598-019-51448-7
14. Boulesnane Y et al (2020) Impact of sampling and DNA extraction methods on skin microbiota assessment. J Microbiol Methods 171:105880. https://doi.org/10.1016/j.mimet.2020.105880
15. Shehadul Islam MAA, Selvaganapathy PR (2017) A review on macroscale and microscale cell lysis methods. Micromachines 8(3):83. https://doi.org/10.3390/mi8030083
16. Marchukov D et al (2023) Benchmarking microbial DNA enrichment protocols from human intestinal biopsies. Front Genet 14:1184473. https://doi.org/10.3389/fgene.2023.1184473
17. Wyatt NJ et al (2025) Evaluation of intestinal biopsy tissue preservation methods to facilitate large-scale mucosal microbiota research. EBioMedicine 112:105550. https://doi.org/10.1016/j.ebiom.2024.105550
18. Eriksson P et al (2017) Evaluation and optimization of microbial DNA extraction from fecal samples of wild Antarctic bird species. Infect Ecol Epidemiol 7(1):1386536. https://doi.org/10.1080/20008686.2017.1386536
19. Yokogawa K et al (1974) Mutanolysin, bacteriolytic agent for cariogenic streptococci: partial purification and properties. Antimicrob Agents Chemother 6(2):156–165. https://doi.org/10.1128/AAC.6.2.156
20. Liu WQ et al (2024) Evaluation of the efficacy of polyethylene glycol in combination with different doses of linaclotide in a fractionated bowel preparation for colonoscopy: a prospective randomized controlled study. Int J Color Dis 39(1):143. https://doi.org/10.1007/s00384-024-04718-4

Chapter 6

16S rRNA Library Preparation and Sequencing

Dave Baker and Cara-Jane Moss

Abstract

This chapter describes the targeting of the prokaryotic 16S ribosomal RNA gene (16S rRNA) using both short-read Illumina Sequencing and long-read Nanopore Sequencing. 16S rRNA targets can be present in multiple copies across the genome and are another consideration when trying to quantitatively determine the proportion of different bacterial species present in samples of different origin, including the gut and feces.

Key words 16S, Ribosome, Short-read, Nanopore, Variable region, Sequencing NGS, PCR, Metabarcoding, Tapestation

1 Introduction

To identify and profile bacterial species of unknown origin in complex samples relies on the identification of a region of the genome that is common or conserved in most if not all species yet contains enough diversity to distinguish one species or genus from another. With the development of NGS (next-generation sequencing) and both Oxford Nanopore and Illumina technology, it is possible to generate millions of reads from a single sample. Also, with more individual species being isolated, cultured, and sequenced individually, this has improved the size and accuracy of reference databases. Here, we describe how to use a DNA sample from a single sample or a mixture of species to generate both long-read nanopore and short-read Illumina libraries for sequencing. In principle, this method can be applied to any conserved target gene, e.g., yeast ITS, animal and fish Co1, or plant RbcL [1]. Here, we present the targeting of the prokaryotic 16S ribosomal RNA gene (16S rRNA) which can be present in multiple copies across the genome.

The 16S rRNA is approximately 1500 bp long and contains nine variable regions surrounded by more highly conserved regions. PCR primers that anneal to the more conserved regions

Simon R. Carding (ed.), *Best Practice in Microbiome Research*, Springer Protocols Handbooks, https://doi.org/10.1007/978-1-0716-5009-7_6,

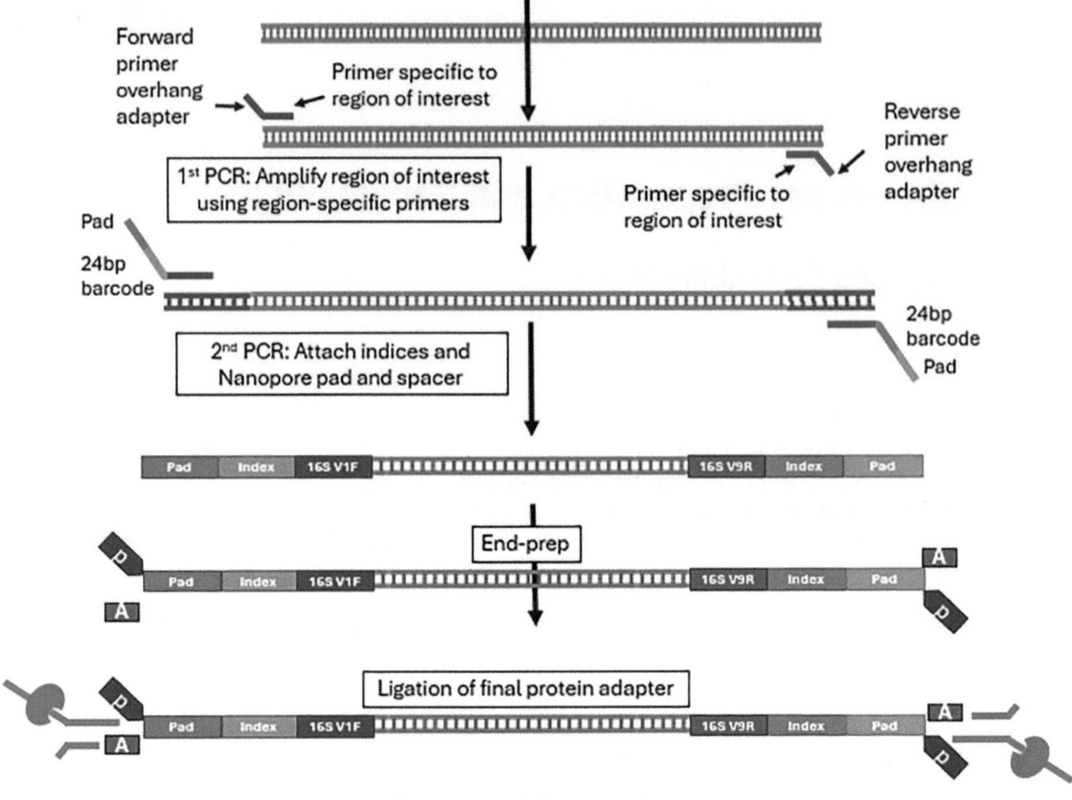

Fig. 1 Full-length 16S for Nanopore sequencing

are used to amplify the variable regions of 16S rRNA enabling identification of genera or species within microbial populations by comparison with well-characterized databases. At the Quadram Institute, we use primer sets for the V1–2, V3–4, and V4 variable regions, for Illumina short-read and full-length V1–9 for Nanopore long-read of the 16S rRNA gene [1]. The general principle of amplification by PCR is shown in Figs. 1 and 2 for Nanopore and Illumina, respectively; also shown is secondary PCR which enables the addition of short sequences (i.e., indexes or barcodes) that enable different samples to be separated, bioinformatically, after a sequencing run.

2 Materials

2.1 Essential Equipment

1. Semi-skirted 96-well plate.
2. Single channel pipettes: 0.1–2.5 μL, 1–10 μL, 5–20 μL, 20–200 μL, 100–1000 μL.

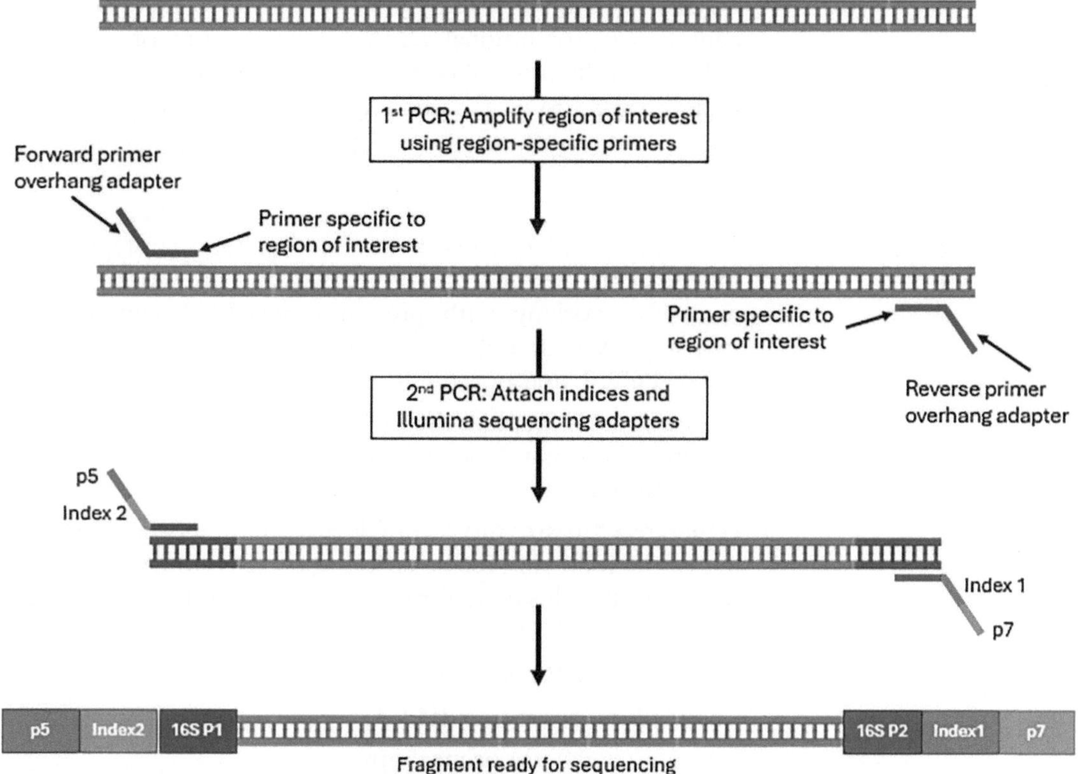

Fig. 2 Short-read 16S for Illumina sequencing

3. Multi-channel pipette 0.5–10 μL.
4. PCR machine.

2.2 Reagents

1. 10 mM Tris-HCl buffer (EB).
2. 80% ethanol (freshly made when needed).
3. Kapa 2G PCR kit (Merck/Sigma).
4. Ampure beads (Kapa or Beckman).
5. Gene-specific primer working stock at 10 μM.
6. Multiple barcoding (UDI's) primers at 10 μM (usually in a 96-well plate).
7. RNAse/DNAse-free dH_2O.

3 Methods

3.1 Diluting DNA to Approximately 5 ng/μL

1. For optimum reproducibility across samples in an experiment, the DNA should be normalized to the same concentration (*see* **Note 1**).

2. DNA is quantified using Qubit or a plate reader. Dilute by adding 5 μL of original DNA to the amount of water n-5 (where n is the original concentration in ng/μL).
3. Seal (*see* **Note 2**) and gently vortex the dilute DNA plate and pulse centrifuge for 5–10 s to collect the liquid at the bottom of the plate wells.

3.2 PCR1 (Gene-Specific PCR)

1. To minimize risk of cross-contamination between steps, physically separate pre- and post-PCR areas and use separate equipment for working with pre- and post-PCR samples. The two-step PCR described here requires three areas: pre-PCR 1, pre-PCR 2, and post-PCR.
2. Before beginning, place the PCR enzyme on ice (not necessary if using "Hot Start" polymerase). Allow all other reagents to reach 20 °C.
3. Generate a "master mix" of PCR reagents and primers according to the following recipe ensuring there is sufficient to fill all sample wells (plus extra for pipetting error and dead volume—usually up to 10%):
   - (i) 10 μL KAPA2G polymerase Master Mix
   - (ii) 7 μL nuclease-free dH2O
   - (iii) 1 μL at 10 μM primer mix.

   When adding or transferring samples or reagent master mixes, change tips between each sample.
4. Add 18 μL of master mix to each reaction well in a PCR plate using a stepper or multichannel pipette from a trough.
5. Add 2 μL of metagenomic DNA from each different sample into an individual well of the PCR plate (*see* **Note 3**).
6. Seal (*see* **Note 2**) the reaction plate with a PCR-grade plate seal ensuring each well is sealed between adjacent wells.
7. Lightly vortex, and spin down gently.
8. Run the first gene-specific PCR (PCR 1) plate on the thermocycler with 95 °C for 3 min, 25 cycles of 95 °C for 10 s, 55 °C for 20 s, and 72 °C for 3 min. For low-biomass samples for the DNA target of interest, increase up to 35 cycles. The extension time covers both full-length and short-read regions of 16S.

3.3 Barcoding: Unique Dual Indexing (PCR 2)

(*see* **Note 4**)

1. During master mix preparation (**step 2**, below), thaw the unique dual indexing (UDI) or CoronaHiT primers for short or long read [2]. Use a different indexed primer plate for each set of 96 samples for the Illumina. The method for full-length 16S only supports barcoding up to 96 samples.

2. Generate a master mix of buffer, DNTPs, and polymerase according to the following recipe ensuring there is sufficient to fill all sample wells (plus extra for pipetting error up to 10%):
   (i) 10 μL KAPA2G Fast HotStart Readymix master mix
   (ii) 8 μL nuclease-free dH2O.
3. Add 1 μL of UDI primers (or CoronaHiT primers [2]) to each well in the new barcoding plate (PCR 2) (*see* **Note 5**).
4. Add 1 μL from each PCR 1 plate well to the corresponding well in the new PCR 2 plate.
5. Add a PCR-grade plate seal (*see* **Note 2**) to the PCR 2 reaction plate.
6. Lightly vortex, and gently centrifuge.

   Run the barcoding plate (PCR 2) on the thermocycler at 72 °C for 3 min, 95 °C for 1 min, ten cycles of 95 °C for 10 s, 55 °C for 20 s, and 72 °C for 3 min.

3.4 Quality Control of Multiplex Barcoding

1. Select two samples per batch in addition to the blanks from each PCR 2 plate and run these on the Tapestation as described above with a D5000 tape with enough spare gel lanes.
2. Label the wells on the Tapestation software for ease of analysis.

3.5 Pooling of Barcoded Multiplexes

1. Quantify the libraries using Qubit or a plate reader. Good quality libraries should be above 10 ng/μL; if less, inspect them further on the Tapestation.
2. In Excel, calculate the μL needed for pooling to ensure that there is sufficient to pool 200 ng for each library. For example, if sample 1 is 50 ng/μL, then pool 200/50 = 4 μL of this library, and so on. Pool into a single tube/Eppendorf.
3. Wash the pooled sample with purification beads (Single SPRI 0.7X) as follows:
   (i) Add 100 μL of pooled sample to a new Eppendorf tube. It is possible to clean the entire volume, although this may be too high a concentration for loading.
   (ii) Add 70 μL of thoroughly mixed fresh beads (this is the 0.7X clean up).
   (iii) Vortex to mix thoroughly.
   (iv) Incubate at 20 °C for 10 min on a Hula mixer. In the absence of a Hula mixer, flick the tube after each 2–3 min of incubation.
   (v) At the end of the 10 min of incubation, place the tube on a magnetic tube rack to sediment the beads which can take several minutes.
   (vi) Remove the supernatant.

(vii) Double-wash the pellet with freshly prepared 80% ethanol. Wait at least 30 s after adding ethanol before removing the supernatant.

(viii) Pulse spin and return to the magnet.

(ix) Remove excess ethanol with a P10 pipette set to 10 μL.

(x) Add 50 μL EB to the sediment, close the lid, and vortex to mix thoroughly.

(xi) Incubate at 20 °C for 5 min on a Hula mixer if available as above.

(xii) Place the tube back on a magnetic tube rack to sediment the purification beads.

(xiii) Transfer the supernatant to a new Eppendorf tube, being careful not to transfer the beads. It is not essential to transfer the entire volume, as it is more important that the beads are not transferred.

(xiv) Label the transferred eluate as "clean library pool" with the date. At this point, the protocol can be halted, and the pool stored at −20 °C.

3.6 Quantification of the Illumina Library

1. Quantify the pooled library using Qubit, using 195 μL of buffer and 5 μL of the pooled library. Note the concentration on the pooled library tube.
2. If short-read Illumina, dilute the pooled library appropriately (using EB) to bring the concentration of the pooled library to ~2 ng/μL and ~ 10 nM) and repeat Qubit (using 195 μL of buffer and 5 μL of the diluted library). For the Nanopore, everything will go into end prep.
3. Once the target concentration has been achieved, run the diluted pooled sample on Tapestation using a D5000 HS reagent and screen tape.
4. On the Tapestation, note the concentration of the final library (this should be approximately the same as the concentration derived by Qubit), the insert size, and the molarity. The insert size will vary depending on which region is amplified. The final library is ready for sequencing. Preparation will vary depending on the instrument you are using.

3.7 Final Nanopore Preparation

1. To 53.5 μL of the pool, add 3.5 μL Ultra II End-prep reaction buffer and 3 μL Ultra II End-prep enzyme mix, flick to mix after each sequential addition.
2. Incubate at 65 °C for 20 min, then at 65 °C for 5 min. Carry out a 1x SPRI bead clean-up with sample purification beads, eluting in 60 μL nuclease-free water.

3. Using Ligation Sequencing Kit V14 (SQK-LSK114), take the amplicon pool and add 5 μL of Ligation Adapter (LA), 25 μL of Ligation Buffer (LNB) and 10 μL of NEBNext Quick T4 DNA Ligase mix and leave at 20 °C for 10 min.
4. Carry out 0.4 X SPRI bead clean-up and wash twice with Short Fragment Buffer (SFB) and elute in 33 μL (13 μL if loading a MinION) of Elution Buffer (EB).
5. Run 1 μL of the pooled sample on the Tapestation using a Genomic reagent and screen tape.
6. Note the concentration of the final library and use a phone app like Promega Biomath to calculate the final loading concentration in fmol. Nanopore advise 5–50 fmol, but empirically, we have found you can load much higher without any adverse effect on the performance of the run or the integrity of the data.

4 Notes

1. For samples of low biomass and/or a large quantity of non-bacterial host DNA present (e.g., skin swabs), this should be increased to increase the number of viable targets. If necessary, determine the relative abundance via qPCR calculate the number of viable targets within the total DNA sample.
2. Include appropriate control samples such as kitome blanks from each process, for instance, a DNA extraction blank, a library prep blank, and if required a ZymoBIOMICS Microbial Community Standard as a positive control.
3. Always seal the 96-well plate with the adhesive seal using a rubber roller or pen to cover the plate and seal between the wells (preferably use foil seals) before the following steps in the protocol:
   (i) Shaking steps (especially at the bead clean-up step).
   (ii) Thermal cycling steps.
   (iii) Centrifuge steps.
4. Barcoding (unique dual indexing; UDI) ensures that libraries sequence and demultiplex with the highest accuracy across all Illumina sequencing platforms. When preparing libraries for multiplexing, Illumina encourages customers to use unique dual indexing (UDI) whenever possible to ensure the most accurate demultiplexing. Every sample will therefore have a unique barcode at both ends which mitigates against "index hopping" which is more prevalent on patterned flow cells to which the DNA attaches for amplification.

5. When adding index adapters with a multichannel pipette, change the tips between each row or column. If using a single-channel pipette, change tips between each sample.

Acknowledgements

This work was funded by the Biotechnology and Biological Sciences Research Council (BBSRC) through BBSRC Core Capability Grants BB/CCG1860/1 and BB/CCG2260/1.

References

1. https://support.illumina.com/content/dam/illumina-support/documents/documentation/chemistry_documentation/16s/16s-metagenomic-library-prep-guide-15044223-b.pdf
2. Baker DJ, Aydin A, Le-Viet T et al (2021) CoronaHiT: high-throughput sequencing of SARS-CoV-2 genomes. Genome Med 13:21. https://doi.org/10.1186/s13073-021-00839-5

Chapter 7

Shotgun Metagenomic Library Preparation

Dave Baker and Cara-Jane Moss

Abstract

This chapter describes the preparation of metagenome DNA samples for Illumina library production and DNA sequencing and, in particular, Shotgun Whole Genome Sequencing (WGS).

Key words PCR, NGS, Illumina, Flow cell, Library, Shotgun, Multiplexed, SPRI, Metagenomics, Microbiome

1 Introduction

Illumina DNA library preparation uses bead tagmentation to randomly insert transposomes into the genomic DNA. This enables amplification and subsequent barcoding of the libraries that can be multiplexed for Illumina sequencing.

This approach is preferred as it is low cost and, when compared across other library construction methods, provides a truer representation of complex samples [1]. Across a range of microbial species with differing % AT base content contained in a mock community, the Flex library construction method outperformed other commonly used library prep, Nextera XT, and was equivalent to more expensive alternatives [2].

2 Materials

2.1 Essential Equipment

1. Semi-skirted 96-well plates.
2. Single channel pipettes 0.1–2.5 μL, 1–10 μL, 5–20 μL, 20–200 μL, 100–1000 μL.

Simon R. Carding (ed.), *Best Practice in Microbiome Research*, Springer Protocols Handbooks, https://doi.org/10.1007/978-1-0716-5009-7_7, © The Author(s) 2026

3. Multi-channel pipette 0.5–10 μL.
4. PCR machine.

2.2 Reagents

1. 10 mM Tris-HCl buffer (EB).
2. 80% ethanol (freshly made when needed).
3. Kapa 2G PCR kit.
4. Nextera Flex Enzyme kit (Illumina® DNA Prep, (M) Tagmentation (96 Samples) 20018705.
5. RNAse-/DNAse-free dH_2O.

3 Methods

The methodology from tagmentation of the genomic DNA and subsequent barcoding PCR is outlined in Fig. 1.

3.1 Diluting DNA to Approximately 5 ng/μL

1. For optimum reproducibility across experimental samples, the DNA should be normalized to the same concentration which is ~5 ng/μL.

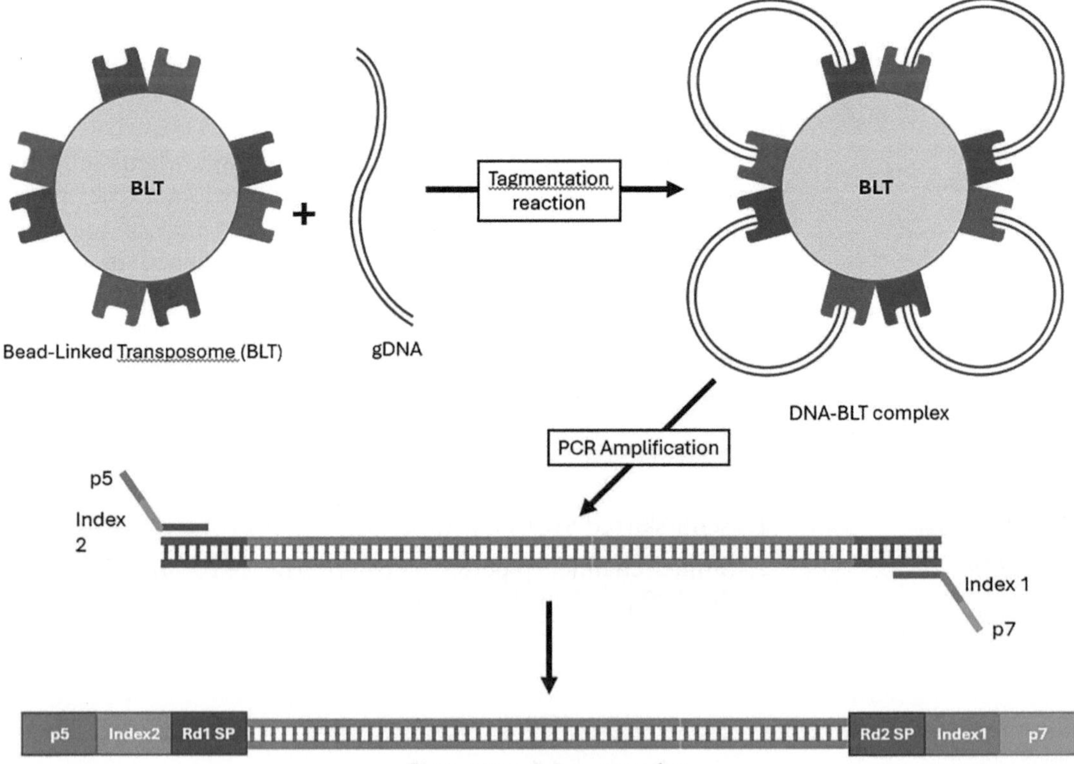

Fig. 1 On-bead tagmentation mediates the simultaneous fragmentation of gDNA and the addition of Illumina sequencing primers. PCR amplifies DNA fragments and adds indexes and adapters

2. DNA is quantified using Qubit or a plate reader. Place 5 μL of extracted DNA suspension into the amount of water x-5 (where x is the concentration in ng/μL). For example, for a sample of 25 ng/μL, add 20 μL (25–5 =) of water to a fresh plate, and then transfer 5 μL of the original DNA stock into this well.
3. Seal and gently vortex the diluted DNA plate and centrifuge briefly (*see* **Note 1**).

3.2 Tagmentation of Metagenomic DNA

1. To avoid the risk of contamination, always physically separate pre- and post-PCR areas and have separate equipment for working with pre- and post-PCR samples. The Quadram Institute has a separate sequencing instrument "dirty" room and is the only place where post-PCR libraries are handled, pooled, and loaded.
2. Keep tagmentation reagents on ice.
3. Ensure sufficient materials are on hand to fill all sample wells (plus extra for pipetting error and dead volume; usually up to 10%). Generate a "master mix" of buffer and tagmentation beads using the following quantities:
   (i) 0.5 μL BLT (bead-linked transposomes).
   (ii) 0.5 μL TB1 (tagmentation buffer 1).
   (iii) 4 μL nuclease-free dH_2O.
   When adding or transferring samples or reagent master mixes, change tips between each sample to prevent contamination.
4. Add 5 μL of this master mix to each reaction well in a new PCR plate (now referred to as the "tagmentation plate") using a stepper or multichannel pipette from a trough.
5. To each well in the tagmentation plate, add 2 μL of metagenomic DNA (one sample per well).
6. Seal the reaction plate with a PCR-grade plate seal to ensure there is no movement of material between the wells.
7. Lightly vortex, and spin down gently.
8. Run the tagmentation plate on the thermocycler at 55 °C for 15 min. Proceed immediately to making the barcoding "master mix" plate.

3.3 Barcoding of Tagmented Samples (See Notes 2–4)

1. Allow the indexed primers (for UDIs) to thaw during "master mix" preparation (below). Use a different indexed primer for every well. UDI plates are laid out in 96-well format, so each sample receives the same corresponding well position barcode.
2. Ensure sufficient materials are on hand to fill all sample wells (plus extra for pipetting error up to 10%), generate a master

mix of buffer, DNTPs, and polymerase according to the following:

(i) 10 μL KAPA2G Fast Hot Start Readymix master mix

(ii) 2 μL nuclease-free dH2O.

3. Add 12 μL of master mix per reaction well in a new PCR plate (now referred to as the barcoding plate).
4. Add 1 μL of indexed primers to each well in the new barcoding plate and use a P10 multichannel pipette so that the index primer pair goes into the corresponding well in the PCR plate.
5. Add 7 μL of multiplex tagmentation reaction and seal the plate using a PCR-grade plate seal.
6. Vortex to mix and spin down gently.
7. Run the barcoding plate on the thermocycler at 72 °C for 3 min, 95 °C for 1 minute, 14 cycles of 95 °C for 10 s, 55 °C for 20 s, and 72 °C for 3 min (*see* **Note 5**).

3.4 Quality Control of Multiplex Barcoding

1. Select two samples per batch in addition to the blanks from each plate and run on the Tapestation as described above on a D5000 tape with enough spare gel lanes.
2. Label the wells on the Tapestation software to ensure ease of analysis.

3.5 Pooling of Barcoded Multiplexes

1. Quantify the libraries using Qubit or a plate reader. Good quality libraries will be between 10 and 50 ng/μL; if less, inspect them further on Tapestation.
2. In Excel, calculate the μL needed for pooling to ensure that there is sufficient to pool 50 ng for each library. For example, if sample 1 is 15 ng/μL, then pool 50/15 = 3.33 μL of this library, and so on. Pool into a single tube/Eppendorf.
3. Wash the pooled sample with sample purification beads (Double SPRI-0.5X-0.7X) as follows:

   (i) Add 96 μL of pooled sample to a new Eppendorf tube.

   (ii) Add 48 μL of thoroughly mixed beads (this is your 0.5X clean up).

   (iii) Vortex to mix thoroughly and pulse spin.

   (iv) Incubate at 20 °C for 10 min on a Hula mixer. In the absence of a Hula mixer, flick the tube to mix after each 2–3 min of incubation.

   (v) At the end of the 10 min incubation, place the tube on a magnetic tube rack to sediment the beads. Wait for the beads to sediment well. This can take up to 5 min.

   (vi) Remove 135 μL of the supernatant and KEEP (transfer to another 1.5 mL Eppendorf).

(vii) Add 18 μL of beads (this is your 0.7X clean up). Vortex to mix and pulse spin.

(viii) Incubate at 20 °C for 10 min on a Hula mixer as above. Place on a magnet, aspirate the supernatant, and discard.

(ix) Double-wash the pellet with freshly prepared 80% ethanol. Wait at least 30 seconds after adding ethanol before aspirating.

(x) Pulse spin and return to the magnet.

(xi) Remove excess ethanol with a P10 pipette set to 10 μL.

(xii) Add 30 μL EB to the sediment and mix thoroughly by vortexing.

(xiii) Incubate at 20 °C for 5 min on a Hula mixer as above.

(xiv) Place the tube back on a magnetic tube rack to sediment the sample purification beads.

(xv) Transfer the supernatant to another Eppendorf tube, being careful not to transfer the beads. Note: It is not essential to transfer the entire volume here, as it is more important to not transfer the beads.

(xvi) Label the transferred eluate as "clean library pool" with the date.

3.6 Quantification of the Illumina Library

1. Quantify the pooled library using Qubit, using 195 μL of buffer and 5 μL of the pooled library.
2. Note the concentration on the pooled library tube.
3. Dilute the pooled library appropriately (using EB) to bring the concentration of the pooled library down to ~2 ng/μL, ~10 nM) and repeat Qubit (using 195 μL of buffer and 5 μL of the diluted library).
4. Once the target concentration is achieved, run the diluted pooled sample on the Tapestation using a D5000 HS reagent and screen tape.
5. On the Tapestation, note the concentration of the final library (this should be approximately the same as the concentration derived by Qubit), the insert size, and the molarity at the ~400–700 bp insert size region.

4 Notes

1. Always seal 96-well plates with an adhesive seal using a rubber roller or pen to cover the plate and seal between the wells before all shaking, thermal cycling, and centrifuge steps.

2. Barcoding of tagmented samples should be carried out immediately after tagmentation.
3. When adding index adapters with a multichannel pipette, change tips between each row or each column. If using a single-channel pipette, change tips between each sample.
4. Alongside other best practices, using unique dual indexing (UDI) ensures that libraries sequence and demultiplex with the highest accuracy across all Illumina sequencing platforms. Thus, when preparing libraries for multiplexing, Illumina encourages customers to use UDIs whenever possible as this ensures that each sample (in every well) will have a unique barcode at both ends. This also mitigates against "index hopping" which is more prevalent in patterned flow cells.
5. For a 20-fold Flex reaction with a total of 10 ng, use 14 cycles of enrichment PCR. This can be reduced to eight cycles, if necessary, but 14 cycles are robust across different template types.

Acknowledgements

This work was funded by the Biotechnology and Biological Sciences Research Council (BBSRC) through BBSRC Core Capability Grants BB/CCG1860/1 and BB/CCG2260/1).

References

1. Baker DJ et al (2021) CoronaHiT: high-throughput sequencing of SARS-CoV-2 genomes. Genome Med 13:21. https://doi.org/10.1186/s13073-021-00839-5
2. Mitsuhiko P et al (2019) Comparison of the sequencing bias of currently available library preparation kits for Illumina sequencing of bacterial genomes and metagenomes. DNA Res 26: 391–398. https://doi.org/10.1093/dnares/dsz017

Chapter 8

Long-Read Metagenomic Sequencing

Dave Baker and Cara-Jane Moss

Abstract

This chapter describes the preparation of metagenomic DNA samples to produce long-read Oxford Nanopore-compatible libraries for shotgun whole-genome sequencing (WGS). This is of value for projects needing to assemble novel unculturable bacterial species and where higher resolution is required.

Key words NGS, Long-read, Nanopore, Flowcell, Library, Shotgun, Native, PCR-free, Metagenomics, Microbiome, Long-read, Nanopore, Full-length, Sequencing NGS, PCR

1 Introduction

This method contains no size selection and is PCR-free and incorporates a damage repair step which is essential if the DNA extraction method used includes bead beating. Nanopore sequencing shares the benefits of short-read Illumina metagenomic shotgun sequencing in addition to enabling assemblies of Metagenome-Assembled Genomes (MAGs) which in turn aids in the discovery and assembly of novel species [1].

2 Materials

2.1 Essential Equipment

1. Semi-skirted 96-well plates.
2. Single-channel pipettes 0.1–2.5, 1–10, 5–20, 20–200, 100–1000 μL.
3. Multichannel pipette 2–10 μL and 10–200 μL.
4. PCR machine.
5. Magnets for tubes and plates.

2.2 Reagents

1. 10 mM Tris-HCl buffer (EB)
2. 80% ethanol (freshly made when needed)

Simon R. Carding (ed.), *Best Practice in Microbiome Research*, Springer Protocols Handbooks, https://doi.org/10.1007/978-1-0716-5009-7_8, © The Author(s) 2026

3. Beckman Ampure beads or equivalent beads.
4. Native Barcoding Kit 96 V14 (Oxford Nanopore).
5. NEB Blunt/TA Ligase Master Mix (NEB).
6. NEBNext FFPE Repair Mix (NEB).
7. NEBNext Ultra II End repair/dA-tailing Module (NEB).
8. NEBNext Quick Ligation Module (NEB).
9. RNAse/DNAse-free dH_2O.

3 Methods

The flow diagram in Fig. 1 provides an overview of the method from DNA damage repair, end repair, barcoded adapter ligation through to the final complex for loading onto a Nanopore flowcell.

3.1 Diluting DNA

1. Normalize the DNA in 48 μL so each sample has an equal amount of DNA (*see* **Notes 1** and **2**).

3.2 DNA Damage and End Repair

1. Thaw and keep the NEB damage and end preparation reagents on ice.

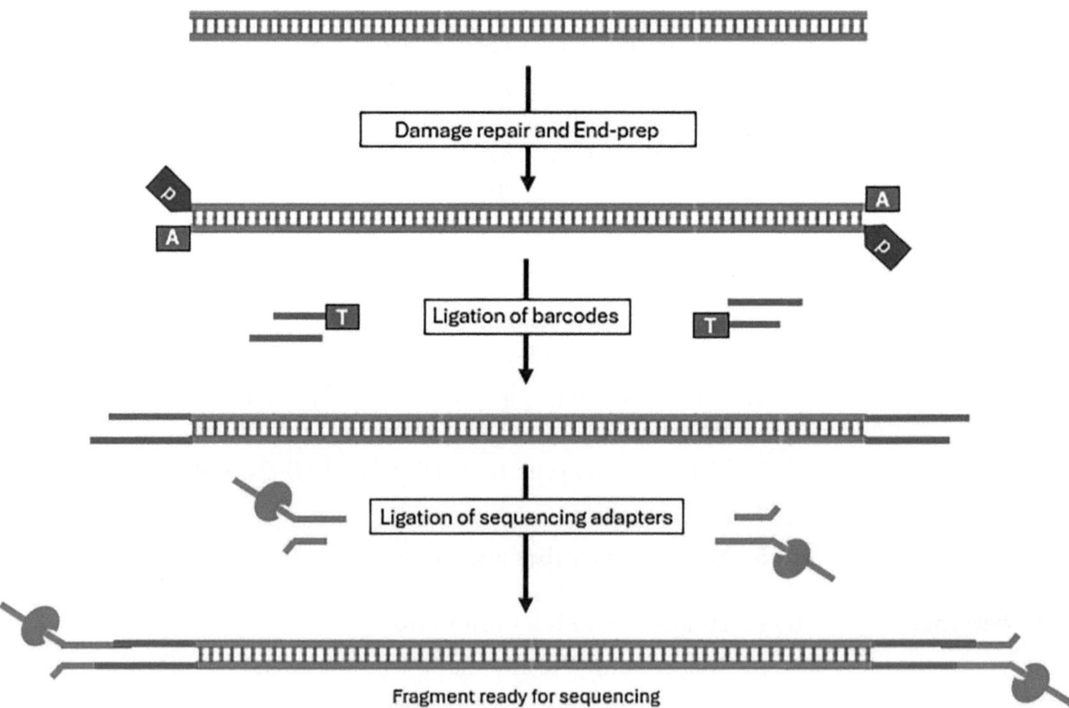

Fig. 1 Nanopore end repair, damage preparation, and Native barcoding followed by final protein adapter libation

2. Ensure sufficient materials are on hand to fill all sample wells (plus extra for pipetting error and dead volume; usually up to 10%). Generate a "master mix" of buffer and enzymes using the following quantities:
   (i) NEBNext FFPE DNA Repair Buffer, 3.5 μL.
   (ii) Ultra II End-prep Reaction Buffer, 3.5 μL.
   (iii) Ultra II End-prep Enzyme Mix, 3 μL.
   (iv) NEBNext FFPE DNA Repair Mix, 2 μL.
3. Add 12 μL of master mix to a fresh 96-well plate.
4. Add the 48 μL normalized DNA to each well.
5. When adding or transferring samples or reagent master mixes, change tips between each sample.
6. Mix by flicking, spin down, and leave at 20 °C for at least 5 min.
7. Place on thermal cycler for another 5 min at 65 °C to deactivate enzymes, cool on ice.
8. To reduce the amount of the Blunt/TA Ligase Master Mix, add an equal volume of Ampure beads and reduce volume by eluting in 10 μL of water. Follow the wash steps in Subheading 3.3.
9. For Native barcode adapter ligation, add 1 μL of adapter from the adapter plate supplied in the library kit, and add an equal amount of Blunt/TA Ligase.
10. Lightly flick, and spin down gently.
11. Leave for 20 min minimum or longer at 20 °C. Seal and store the plate at 4°C. for 16 hours if necessary (*see* **Note 3**).

3.3 Clean Up Adapter-Ligated Samples

1. Nanopore deactivation of ligase uses EDTA. However, to eliminate the presence of any residual active ligase and free end and adapters after pooling, a 1× ampure bead clean and elution in 10 μL of water is recommended.
2. Once resuspended, centrifuge and incubate for 5 min at 20 °C.
3. Place on the magnet to remove the entire volume of each well into a single tube.
4. Perform a further clean-up step to reduce the volume to 30 μL in water (at this point, the pool can be stored at 4°C for up to a few weeks), ready for final native adapter ligation in preparation for loading.

3.4 Final Adapter Ligation Prior to Loading (See Notes 4 and 5)

1. In a 1.5-mL Eppendorf LoBind tube, mix in the following order (flick mix and pulse spin in mini centrifuge between each step):
   (i) Pooled barcoded sample, 30 μL.
   (ii) Native Adapter (NA), 5 μL.

(iii) NEBNext Quick Ligation Reaction Buffer (5×), 10 μL.

(iv) Quick T4 DNA Ligase, 5 μL.

(v) Total = 50 μL.

2. Incubate at 20°C for a minimum of 20 min.
3. Carry out 0.4× Ampure bead clean-up and wash twice with short fragment buffer (SFB) removing the tube from the magnet during the wash before returning to the magnet. Finally, elute in 33 μL (13 μL if loading a MinION) of elution buffer (EB).

3.5 Quantification and Loading of the Final, Nanopore Pool

1. Run 1 μL of the pooled sample on the TapeStation using a Genomic reagent and screen tape.
2. On the TapeStation, note the concentration of the final library and use a phone app like Promega Biomath to calculate the final loading concentration in fmol. Nanopore advises 5–50 fmol, but empirically, we have found you can load much higher without any adverse effect on the performance of the run or the integrity of the data.

4 Notes

1. Typically, 5–50 fmol of final library is recommended for loading on a MinION or PromethION cell. However, with kit 14, R10, there are no issues with overloading with no need to re-fuel. Due to the average yield from a cell being variable, between 8 and 12 samples are added on a PromethION cell to achieve approximately 10Gbp of data for each sample. During library construction, there are losses, but a good starting point is using at least 1 μg of DNA per sample. If all samples are fragmented in a similar manner, then normalizing to a set concentration and volume is ideal.
2. With high molecular weight (HMW) DNA, there should be minimal or no vortexing. However, with samples extracted mechanically, fragmentation can occur and these sample types usually have a DNA size that is unaffected by a vortex; however, always treat DNA samples with care for any long-read sequencing platforms and applications and flick mix for all mixing steps.
3. Always seal 96-well plates with an adhesive seal using a rubber roller or pen to cover the plate and seal between the wells before all shaking, thermal cycling, and centrifuge steps.
4. As Native barcoding is PCR-free, there is no need to separate pre- and post-PCR as for Illumina shotgun sequencing.
5. When adding index adapters with a multichannel pipette, change the tips between each row or column. If using a single-channel pipette, change tips between each sample.

Acknowledgements

This work was funded by the Biotechnology and Biological Sciences Research Council (BBSRC) through BBSRC Core Capability Grants BB/CCG1860/1 and BB/CCG2260/1.

Reference

1. Jain M et al (2016) The Oxford Nanopore MinION: delivery of nanopore sequencing to the genomics community. Genome Biol 17:239. https://doi.org/10.1186/s13059-016-1103-0

Chapter 9

Reproducible and Scalable Bioinformatics

Samuel J. Haynes, Alise J. Ponsero, Viet Thanh Le, and Andrea Telatin

Abstract

Modern microbiome research requires complex bioinformatics workflows that process large volumes of sequencing data through multiple steps. The reproducibility of these analyses remains a significant challenge due to the complexity of software dependencies, computational requirements, and the diversity of analytical tools. This chapter presents fundamental concepts and best practices for creating reproducible and scalable bioinformatics workflows, with a particular focus on microbiome data analysis. We introduce infrastructure considerations for computational analyses; discuss modern solutions for managing software dependencies; demonstrate how version control and comprehensive documentation ensure analytical transparency and reproducibility, aligning with FAIR (Findable, Accessible, Interoperable, and Reusable) principles; and emphasize practical solutions in transitioning from ad-hoc scripts to robust, reproducible workflows that can handle datasets of any size while maintaining computational efficiency and analytical integrity.

Key words Bioinformatics, Scalable, Transparency, Reproducibility, FAIR principles

1 Introduction

The analysis of microbiome data begins with raw sequencing reads stored in FASTQ format files, which typically contain thousands to millions of short DNA sequences. These reads must undergo several processing steps including quality control, denoising or clustering, and taxonomic assignment to generate interpretable ecological data matrices. This data reduction process transforms raw sequences into features that can be counted across samples, enabling downstream statistical analyses and biological interpretation.

Each step in this analytical pipeline requires specialized software tools that are frequently updated to incorporate new algorithms, fix bugs, or handle new data types. The same applies to reference databases, updated to include newly described organisms. This constant evolution of tools and databases, while beneficial for the field, poses significant challenges for analysis reproducibility.

Simon R. Carding (ed.), *Best Practice in Microbiome Research*, Springer Protocols Handbooks,
https://doi.org/10.1007/978-1-0716-5009-7_9,

Containerization technologies like Docker and Singularity have emerged as a solution to ensure reproducible analyses by packaging software tools and their dependencies into isolated, portable units. Containers effectively create snapshots of the exact software versions used in an analysis, allowing researchers to reproduce results consistently across different computing environments and time points. This approach addresses a common challenge in bioinformatics where analyses may fail to reproduce due to version conflicts or missing dependencies.

Microbiome datasets of increasing size and complexity require scalable computing. Two approaches are institutional high-performance computing (HPC) clusters and cloud computing platforms [1, 2]. HPC clusters provide dedicated resources and are typically accessed through job scheduling systems, while cloud platforms like Amazon AWS or Google Cloud offer flexible, on-demand computing resources [3, 4]. Each approach has distinct advantages. HPC clusters often provide cost-effective access to large computing resources for institutional users, while cloud platforms offer flexibility and eliminate the need for local infrastructure maintenance.

Workflow managers like Nextflow bridge the gap between reproducibility and scalability by providing a framework to orchestrate complex analytical pipelines. Nextflow allows researchers to define their analysis as a series of interconnected processes, each potentially running in its own container. A key advantage of Nextflow is its portability. The same workflow can be executed seamlessly across different computing environments, from personal laptops to HPC clusters or cloud platforms, while maintaining reproducibility through containerization. This flexibility enables researchers to develop and test workflows locally before scaling up to larger computational resources as needed.

2 Methods

2.1 Where to Perform Bioinformatics Analyses

The **Command Line Interface (CLI)** is an essential tool for conducting bioinformatics analyses across different computing environments [5, 6]. Linux and macOS come with a native terminal interface, and recent versions of Windows (since October 2017) can install the Windows Subsystem for Linux (WSL), which offers a Linux-compatible CLI.

The CLI is a text-based interface where the user can type commands, which are executed, and each command can print more text on the terminal. This simple concept allows the user to combine different tools in a coherent workspace.

For example, the top BLAST hits can be extracted for a sequence called "Seq1" using a command like (*see* **Note 1** for details):

```
awk '$1=="Seq1"' blast_results.tsv | sort -k11,11n -g | head -n
5
```

2.1.1 The UNIX Philosophy

The popularity of Linux as the workbench for bioinformatics analysis is based on a set of norms called the "Unix philosophy of programming" initially described by Ken Thompson [7]. Two key principles are as follows:

1. Modularity and specificity: Each computational tool should perform a single, well-defined function with high precision. Rather than developing monolithic applications that attempt multiple analytical tasks, create specialized tools for each distinct operation in the analytical pipeline.
2. Interoperability: Design tools to process standardized input and generate output that can serve as input for subsequent analytical steps. This principle facilitates the construction of flexible analytical pipelines where tools can be interchanged or modified without disrupting the overall workflow. For instance, FASTQ and FASTA formats serve as universal intermediates in sequence analysis pipelines, enabling seamless data flow between different tools.

A bioinformatic protocol is sometimes facilitated by these design principles, as the output of a tool can be fed into another tool as input, for example, sorting alignment by coordinates.

There are several alignment tools, and they adopt a common format to store alignments (the SAM format). A package called *samtools* deals with the SAM format and has a function to convert it to the more efficient binary format (BAM), and another function to sort it. This means that aligners are not required to implement the conversion and sorting parts (*see* **Note 2**):

```
minimap ref.fasta reads.fq.gz | \
 samtools view -b | \
 samtools sort -o sort.bam -
```

While it is not always possible to directly chain bioinformatics tools in this way, a bioinformatics workflow is a combination of several tools.

2.1.2 HPC and Cloud

While basic analyses can be performed on local machines, the increasing size and complexity of genomic datasets often require more powerful computing resources. Modern bioinformaticians typically work across three main environments:

1. Local workstations for development and small-scale analyses (e.g., statistical analysis using R).

2. HPC clusters.
3. Public cloud providers for large-scale genomic processing pipelines (see **Note 3** for details).

The choice of computing environment depends on factors like data volume, computational requirements, and time constraints, with many institutions providing access to a combination of these resources to support different analytical needs.

2.2 Managing Software Dependencies

Bioinformatics protocols rely on multiple software packages, each built upon specific versions of existing tools and libraries (reusable components of software tools), collectively referred to as dependencies. A common challenge arises when different tools in a protocol require conflicting versions of the same dependency. Version differences can significantly impact analysis outcomes, whether due to bug fixes or changes in default parameters.

A broadly adopted method to address these dependency challenges is the adoption of the package manager Conda, combined with the community-driven repository Bioconda channel [8]. Bioconda (https://bioconda.github.io) maintains thousands of bioinformatics packages with predefined dependencies, enabling users to create isolated environments where specific software versions coexist without conflicts. For example, to create an environment for microbiome analysis, you can specify exact versions in a single command:

```
conda create -n microbiome "cutadapt>=1.3" bioconductor-
dada2=1.24
```

This command will create a new working environment called "microbiome" containing c*utadapt* (we request a version equal to or above 1.3) and *DADA2* (version 1.24), both being packages used in microbiome analysis. Since DADA2 is an R library, Conda will also install R itself. So, our environment will contain a program to remove primers and the DADA2 package.

These environments can be shared with collaborators through an *environment.yml* file, which lists all the packages that Conda installed in the environment, not only the ones explicitly requested but also all the dependencies, ensuring identical software configurations across different systems. Conda environments are particularly useful when developing new pipelines or easily installing new packages.

2.2.1 Micromamba Is a Faster Alternative to Conda

Micromamba (https://mamba.readthedocs.io/en/latest/user_guide/micromamba.html) is a reimplementation of the core features of Conda with faster installation and environment creation times. Complex bioinformatics environments can be very difficult to implement for Conda, taking lots of time, or can even fail.

Micromamba overcomes Conda's limitations through a streamlined C++ codebase and improved dependency resolution algorithms. Micromamba maintains full compatibility with existing conda channels and environment files, making it a drop-in replacement. The key difference lies in the command syntax—instead of `conda create`, users type `micromamba create`. For instance, the previous environment would be created using:

```
micromamba create -n microbiome "cutadapt>=1.3" bioconductor-dada2=1.24
```

This faster alternative is particularly valuable when working with complex bioinformatics environments that contain many interconnected packages, potentially reducing setup time from minutes to seconds.

Micromamba simplifies the installation of bioinformatics packages, but recreating environments over time poses challenges. Even when specifying exact package versions, their dependencies may update independently, leading to inconsistent environments. While exporting the complete environment to a YAML file (using `micromamba env export`) captures all package versions and dependencies, this method has its limitations. Security updates may remove older library versions, preventing successful environment recreation from the YAML file later. To be able to run the same environment in the future, a more robust way of freezing an environment is needed.

2.3 Container Technologies Provide Robust Dependency Management

Containers are self-contained computing environments that package software applications, their dependencies, and a minimal operating system into a single, portable unit. A container is a standardized, isolated environment that contains everything needed to run a specific set of programs.

Docker and Apptainer (formerly Singularity) are two leading container technologies in bioinformatics. While both achieve similar goals, they serve different use cases. Docker is prevalent in development environments and cloud computing, offering extensive configuration options and a rich ecosystem of tools. Apptainer, with its enhanced security model and user-space implementation, is better suited for HPC environments where system administrators prioritize security and multi-user management.

For microbiome analysis, containers ensure computational reproducibility by maintaining consistent software environments across different computing systems. When sharing a container, the exact computational environment used for your analysis is also shared. This is particularly valuable for complex bioinformatics pipelines that depend on multiple software tools and specific versions.

The Conda/Micromamba ecosystem integrates well with container development and can install and manage bioinformatics packages within individual containers during the build process. This approach combines Bioconda's extensive collection of bioinformatics packages with the portability of containers. For example, a Dockerfile can be created that uses Conda to install a specific version of tools like QIIME2 (used in metabarcoding analysis), ensuring these tools and their dependencies are permanently captured in the container image.

Bioinformatics containers are maintained through centralized repositories like Biocontainers (https://biocontainers.pro), which provide version-controlled tools with documented dependencies for microbiome analysis workflows [9].

Note: Once a package is added to the Bioconda channel, all the new releases of the same package are automatically added. The Galaxy project [10] will generate Apptainer (Singularity) containers automatically for all the tools in Bioconda, and the Biocontainers project will make Docker containers available. In this way, externally available containers can be relied on if the workflow is split into atomic steps, each using a single container for a specific tool.

3 Ensuring Version Control and Documentation

Reproducible analyses demand findable, accessible, interoperable, and reusable (FAIR) code [11]. Publishing data following FAIR principles is a requirement for submitting scientific articles, and FAIR principles for research software are gaining traction. Software contributions to research are gradually being acknowledged as a valid contribution, so ensuring code is openly accessible enables you to get credit for your work and provides a source for citation.

Publishing analysis code in a findable and accessible repository is the simplest step in achieving FAIR research software. Zenodo is a generic research data repository hosted by the CERN data center. Uploading analysis scripts to Zenodo with descriptive metadata provides an immutable and permanent domain object identifier (DOI) that can be added to publications. However, using a code development platform such as GitHub [12] or GitLab to host published code can help ensure it is also interoperable and reusable.

GitHub and its peers provide tools that support best practices in code development prior to publication. Issue tickets, maintain notes throughout the development process, and collect responses from the development team and software users. Issues are not limited to solving bugs. Labels can be assigned to differentiate tickets that discuss bugs, possible enhancements, and calls for help. Issue tickets can be easily integrated into project-level kanban boards to facilitate higher-level team management and task prioritization. Solutions to issues are controlled and evaluated through pull requests. This allows solutions to be submitted by anyone worldwide with any proposed changes to the code highlighted together with a brief justification. Members of the development team then evaluate the request and decide whether to pull the solution into the repository. Template pull requests and issue tickets can be created to aid contributors in providing sufficient details for the requests to be evaluated.

Development platforms support software after publication by hosting documentation, automating the integration of updates, and separating software versions. Most platforms automatically render markdown `.md` files as HTML that present an overview of the software within the repository in a reader-friendly manner, often as `README.md`. Detailed software manuals can be hosted within wikis on the repo or documentation websites which can be generated and hosted by GitHub pages. Additional functionality and bug patches can be automatically checked and accepted into the code using continuous integration tools. GitHub Actions provides an API to run test scripts on every new pull request and automatically accept them if they pass. Actions can be configured to test code on different OS types and versions. Code changes can be grouped together to form minor updates, and major updates can be assigned specific versions that users can specify when they download to avoid legacy issues. Furthermore, GitHub automatically tracks who contributed to each code change so you can ensure everyone gets sufficient credit.

3.1 Managing Sequencing Datasets

The output of NGS sequencing (see Chapter 8) consists of a set of files in FASTQ format. While most output files of bioinformatics analysis can be reproduced, the raw output of sequencing is a primary data, hence particularly precious. For secure storage, it's recommended to have a sequencing output management system. The storage should be redundant, so that physical failure of a single hard disk will not impact the integrity of the files, and possibly spread across multiple data centers (to mitigate the risk of destruction by fire or water accidents).

In addition to a secure storage location, a data management system ensures control of metadata and centralized and controlled access to the datasets. The IRIDA (Integrated Rapid Infectious Disease Analysis) platform [13] is an open-source bioinformatics framework that can be hosted in Linux servers or virtual machines, allows management of users with controlled access to different datasets, and provides different ways to upload raw data into the system (from the command line, or with dedicated tools) and to retrieve data (again from the command line, or with connectors to other platform including Galaxy [10]), as summarized in Fig. 1.

This user-friendly web-based platform provides an end-to-end solution encompassing data management, analysis pipelines, and secure sharing mechanisms.

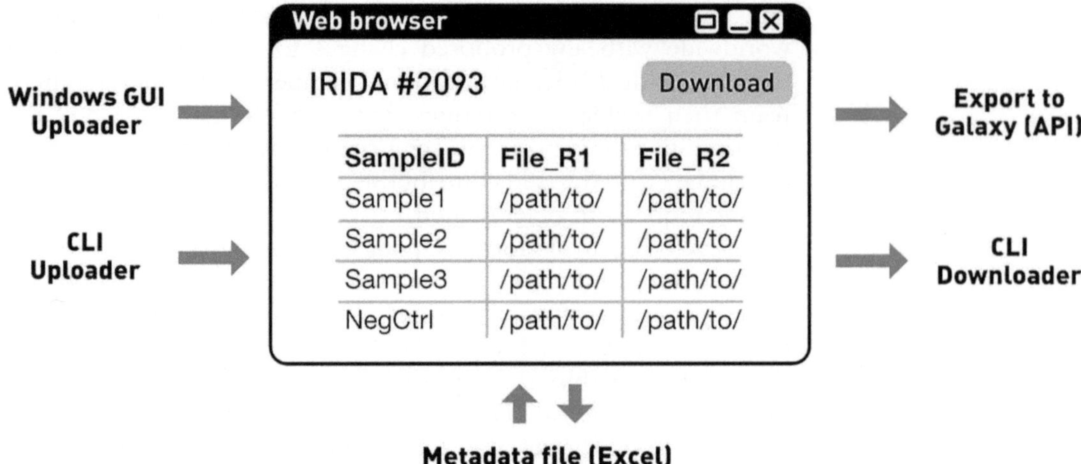

Fig. 1 IRIDA is a web-based application with multiple interfaces to upload and retrieve data. From the web browser, it is possible to download and upload (to update) the metadata (any set of attributes of each sample)

Fig. 2 Schematic representation of a bioinformatics workflow. Arrows represent key processing steps, showing both the analytical operation (above) and commonly used bioinformatics tools (below). This simplified example illustrates the sequential transformation of data through the pipeline

This architectural design, combined with rigorous adherence to public health best practices and FAIR data principles, positions IRIDA as a transformative tool in modernizing infectious disease surveillance and outbreak response within multi-jurisdictional public health networks.

An interesting feature of IRIDA is the ability to launch analytical pipelines directly from the web interface.

4 Orchestrating Bioinformatics Workflows

Bioinformatics analyses involve several interconnected steps, defined as workflows in computer science and equivalent to a laboratory protocol with a minimal example provided in Fig. 2.

A workflow is scalable, and if the number of samples to process increases, a way to parallelize the processing is needed. For smaller datasets, the analysis can be performed on a local computer, with multiple samples needing a suitable infrastructure, commonly an HPC or a cloud environment with very large studies needing public cloud services (such as AWS from Amazon or Azure from

Microsoft) which can be rented for the time needed for the execution of the workflow.

For each step, it needs to be ensured that:

1. It can access the required tool.
2. It can access the appropriate resources (some steps are memory-intensive, like the de novo assembly in the example, while others will benefit from multiple processing threads but less memory).
3. A downstream step only starts when the feeding step has successfully finished and produced the expected output (which will be the input of the following step).

Two popular workflow languages are Nextflow (a domain-specific language based on the less popular programming language Groovy) and Snakemake (based on the Python programming language) [14, 15]. They both allow the orchestration of tasks and ensure reproducible execution of complex workflows, and they are both actively maintained. Nextflow is currently the only one offering both professional support through the Sequera company (https://seqera.io/) and a community-driven effort in sharing best-practice workflow called nf-core [16].

4.1 Nextflow: A Workflow Language and Manager

Nextflow is both a domain-specific language (DSL) and a workflow management system designed for complex scientific pipelines. Built on Groovy, it provides a declarative syntax for defining computational workflows as a series of processes, each representing a single analytical step. The power of Nextflow lies in its ability to abstract the execution layer from the workflow definition with the same pipeline running seamlessly on a local computer, an institutional cluster, or cloud infrastructure without modifying the workflow code.

Nextflow achieves this scalability through its execution model. Each process in a workflow runs in its own isolated environment, with Nextflow automatically managing data flow between processes. The system monitors resource usage will retry execution on failure and provides detailed execution reports.

Container integration is central to Nextflow's reproducibility strategy. Each process can specify a container image (Docker or Apptainer), ensuring that the software environment remains consistent across different computing platforms. The container directives in Nextflow are straightforward with a single line specifying the container image ensuring that the process executes within that containerized environment. This seamless integration ensures complex software dependencies are handled transparently, while maintaining reproducibility.

Nextflow's integration with GitHub streamlines pipeline sharing and versioning. For example, the following command:

```
nextflow run nf-core/ampliseq --input ex.csv --outdir data
-profile docker
```

will first fetch the open-source pipeline "ampliseq" from GitHub, then will use Docker for dependency management and will autonomously fetch all the required containers. Finally, it will orchestrate all the tasks as instructed by our Nextflow configuration: if using an HPC with the Slurm scheduler, Nextflow will submit to the appropriate queues all the tasks.

Nextflow also supports version control through git tags, allowing users to run specific versions of a pipeline, enhancing reproducibility. The nf-core community maintains a collection of peer-reviewed, standardized pipelines that serve as both ready-to-use tools and educational resources for pipeline development. For a first overview of Nextflow, see the training website https://training.nextflow.io/, while for a catalog of curated pipelines by the nf-core community, see https://nf-co.re/.

For the collection of protocols in this book, the nf-core/ampliseq (for metabarcoding analysis), nf-core/taxprofiler (for reads-based taxonomy profiling of whole metagenome shotgun experiments), and nf-core/mag (for assembly and binning of metagenome shotgun) are particularly relevant (see Chapter 8).

5 Common File Formats

5.1 Standard Files Used in CLI

Plain text files play a pivotal role in the interchange of information among CLI programs [17], and several bioinformatics files are plain text files adopting a definite structure.

5.1.1 FASTA Format

This represents the simplest structure, consisting of a description line starting with ">" followed by sequence data in subsequent lines, supporting both nucleotide and protein sequences. FASTQ extends this concept by incorporating quality scores for each base (the probability that the base was called wrongly while processing the raw data from the sequencer), using a four-line structure per sequence: the sequence identifier with an "@" prefix, the raw sequence, a "+" separator line, and ASCII-encoded quality values. Utilities such as SeqKit and SeqFu allow comprehensive manipulation of FASTA and FASTQ files [18, 19], while fastp focuses on filtering FASTQ files based on various criteria [20].

5.1.2 BED (Browser Extensible Data) Format

BED is suitable for genomic features and annotations, the specifies genomic intervals using tab-separated fields, with the first three mandatory columns defining chromosome, start, and end positions [21]. The program *bedtools* is a suite of utilities to manipulate BED files [22].

5.1.3 GFF3 (General Feature Format Version 3)

GFF3 provides a more extensive structured approach to describing genomic features, utilizing nine required fields including sequence ID, source, type, coordinates, and attributes, making it particularly suitable for genome annotation projects. Some of the tools for parsing GFF are Gff Utils and AGAT [22, 23].

5.1.4 Newick Format

This represents evolutionary relationships between organisms as a text-based tree structure using nested parentheses and commas, where branch lengths can be specified with colons and each branch leads to either another set of branches or ends in a tip (leaf) labelled with a species or sequence name. The Newick utilities are command-line tools to manipulate phylogenetic trees stored in Newick format [24].

5.1.5 SAM (Sequence Alignment/Map) Format

Data in this format begins with header lines (prefixed with "@") followed by alignment records containing essential information such as read names, mapping positions, and details describing the alignment in terms of insertions and deletions. This format is supported by a suite of tools called *samtools* that allow conversion and manipulation of SAM files [25].

All these file formats are based on plain text files, and an example of their structure and their interconnection is summarized in Fig. 3.

5.2 Formats to Describe Microbiome Studies

In microbiome studies, there is a broad use of tabular files (TSV, for tab-separated values, or CSV, for comma-separated values) to store metadata, count tables, and taxonomy annotations, but without a proper standard. A notable exception is the BIOM (Biological Observation Matrix) format which is a JSON-based file structure designed to efficiently store biological samples by observation contingency tables along with their associated sample and observation metadata, making it particularly valuable for representing microbiome abundance matrices and their contextual information in metagenomics studies [26].

A set of samples can be described by a mapping file (sometimes called a metadata file) that associates an identifier for each sample with a set of attributes. After analyzing the sequencing output, a feature table (or count table) reports for each sample the raw abundance of each taxon. In the case of metabarcoding experiments, the count table will report the raw abundance of each representative sequence, with a taxonomy file mapping the predicted taxonomy assignment for each sequence, and finally, a sequence tree can report the similarity between features (Fig. 4). Mapping file, count table, and taxonomy can be embedded in a single BIOM file, making this format very useful, despite its relatively low adoption by bioinformatics tools.

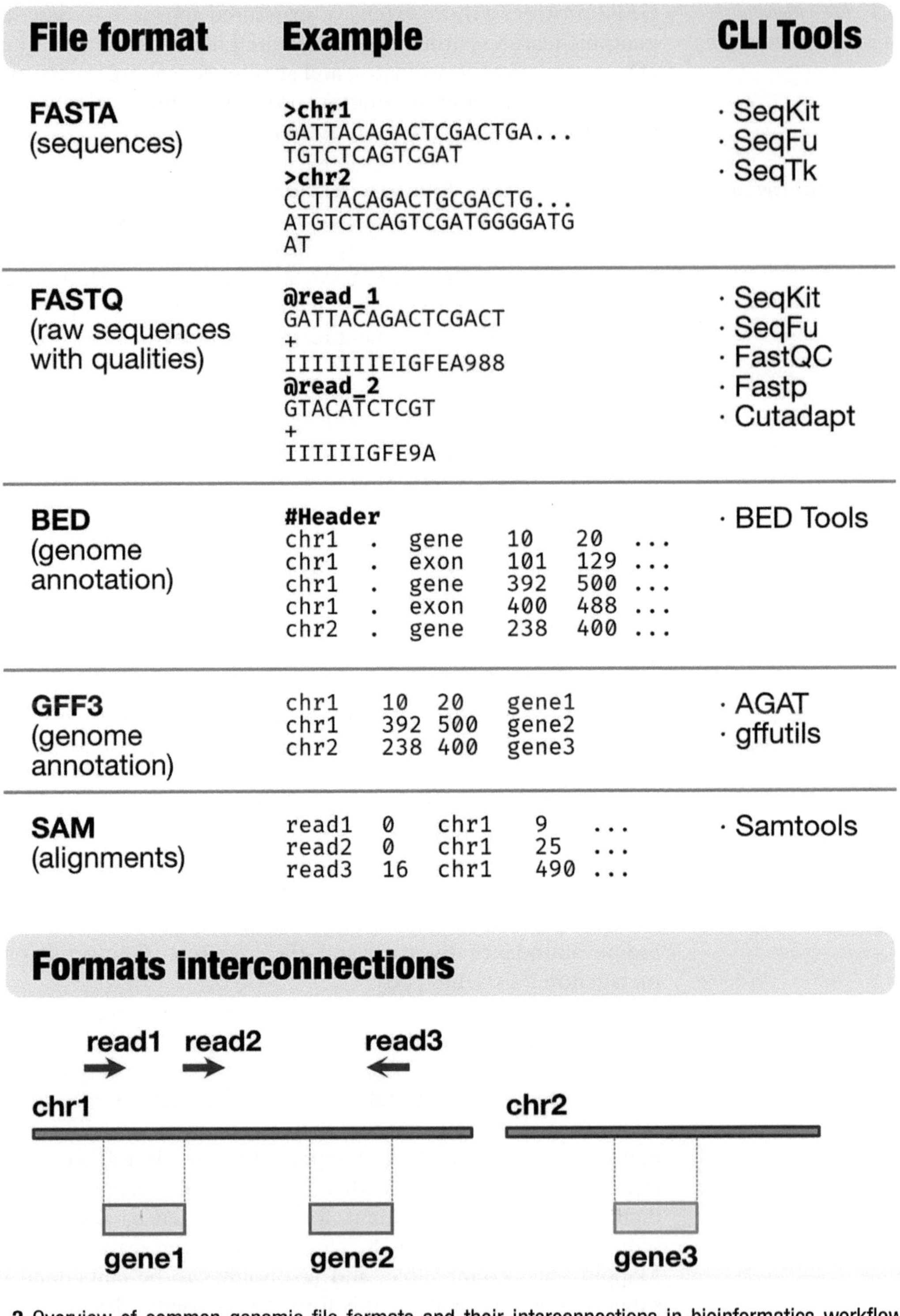

File format	Example	CLI Tools
FASTA (sequences)	**>chr1** GATTACAGACTCGACTGA... TGTCTCAGTCGAT **>chr2** CCTTACAGACTGCGACTG... ATGTCTCAGTCGATGGGGATG AT	· SeqKit · SeqFu · SeqTk
FASTQ (raw sequences with qualities)	**@read_1** GATTACAGACTCGACT + IIIIIIIEIGFEA988 **@read_2** GTACATCTCGT + IIIIIIGFE9A	· SeqKit · SeqFu · FastQC · Fastp · Cutadapt
BED (genome annotation)	**#Header** chr1 . gene 10 20 ... chr1 . exon 101 129 ... chr1 . gene 392 500 ... chr1 . exon 400 488 ... chr2 . gene 238 400 ...	· BED Tools
GFF3 (genome annotation)	chr1 10 20 gene1 chr1 392 500 gene2 chr2 238 400 gene3	· AGAT · gffutils
SAM (alignments)	read1 0 chr1 9 ... read2 0 chr1 25 ... read3 16 chr1 490 ...	· Samtools

Formats interconnections

read1 read2 read3

chr1 chr2

gene1 gene2 gene3

Fig. 3 Overview of common genomic file formats and their interconnections in bioinformatics workflows, illustrating the standard data structures for sequence data (FASTA, FASTQ), genome annotations (BED, GFF3), and sequence alignments (SAM). The lower panel demonstrates how these formats integrate in a typical analysis pipeline, showing the spatial relationships between a reference sequence (chr1, chr2) in FASTA format, aligned reads (read1, read2, read3) that are sequences in FASTQ format and displayed in the location

5.3 Importing Microbiome Datasets in R

PhyloSeq is an R package designed to facilitate the analysis and visualization of microbiome census data. The package integrates the core data types produced by high-throughput microbiome sequencing: abundance tables (OTU or ASV counts), taxonomic assignments, sample metadata, and phylogenetic trees [27]. By combining these elements into a unified data structure called a "phyloseq object," the package streamlines complex microbiome analyses without requiring users to manually maintain multiple data files or handle format conversions. PhyloSeq provides extensive functions for data import from various sources (including BIOM files and common sequencing platforms), data preprocessing (filtering, normalization, transformation), diversity analysis (alpha and beta diversity metrics), and publication-quality visualization using ggplot2. The package's integration with other popular R packages in the microbiome analysis ecosystem, such as vegan [28] (multivariate analysis of ecological communities) and DESeq2 [29] (differential abundance analysis of count data), makes it a central tool in microbiome research workflows, particularly for statistical analysis and the creation of publication-ready figures.

5.4 Qiime2 Artifacts

Qiime2 is a widely adopted framework to analyze metabarcoding datasets [30] consisting of a set of coherent interfaces around third-party tools enabling the end-to-end analysis of raw sequencing reads. All input and output files are encapsulated in compressed archives (zip) called *artifacts*. An artifact is a file (FASTA, Newick, table, etc.) coupled with metadata, bibliography, and provenance. Given a single *artifact* (for example, a count table), it is possible to investigate all the steps performed to produce it, including the parameters used to perform each step. While this brings a practical improvement in traceability and reproducibility, only tools encapsulated in Qiime2 can adopt this. A command line utility called *qax* allows to inspect artifacts without having Qiime2 installed [31].

While Qiime2 can be used in whole metagenome sequencing datasets and for taxonomic profiling of the reads [32] (see Chapter 8), its adoption is very limited and its use is not therefore recommended for these datasets.

Fig. 3 (continued) they map to thanks to a SAM file. Three genes are shown, and their coordinates can be stored in the generic BED file or in a more gene-centered GFF3 file. Associated bioinformatics tools (SeqKit, FastQC, Samtools, etc.) are indicated for each format, representing the standard computational toolkit for genomic data processing

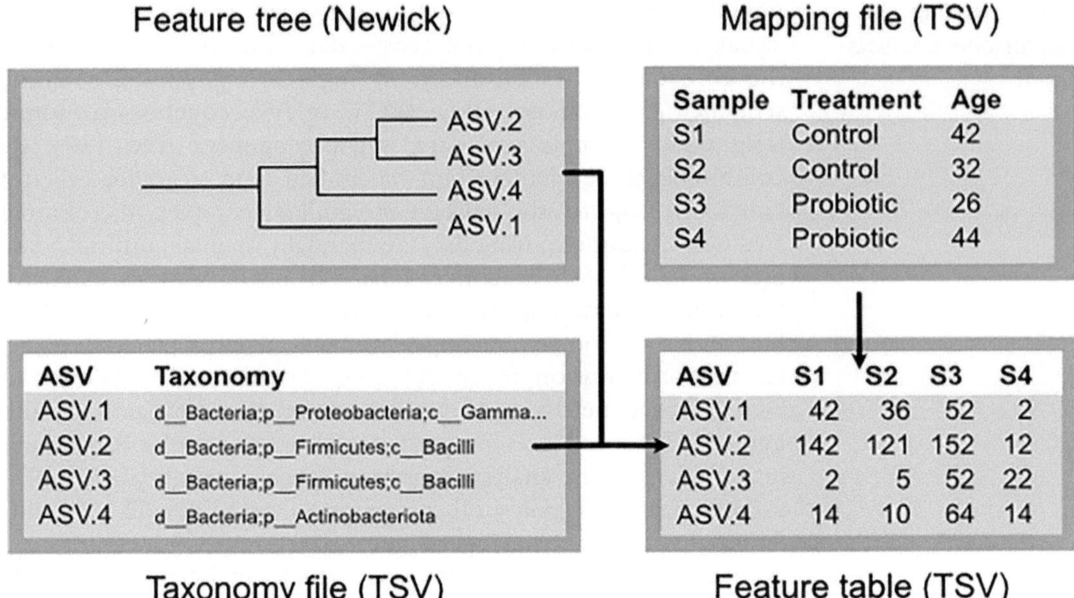

Fig. 4 Files typically produced from a metabarcoding experiment. Four key files are produced that provide complementary information: (1) a Newick-formatted feature tree showing the phylogenetic relationships between ASVs (Amplicon Sequence Variants), (2) a mapping file in TSV format containing sample metadata including treatment conditions and subject age, (3) a taxonomy file associating each ASV with its taxonomic classification from domain to genus level, and (4) a feature table displaying the abundance of each ASV across different samples. Together, these files enable a comprehensive analysis of microbial community composition, structure, and metadata associations

6 Notes

1. This command works by combining multiple programs via the pipe operator (the symbol "|"). First, awk extracts the lines where the first column matches the string OTU1, then we sort these rows numerically using the value in the 11th column, and finally, we only print the first five records (with head). These three commands (awk, sort, and head) are commonly present in any Linux system, but often we will need to install dedicated bioinformatics programs (e.g., BLAST, used to generate the blast_results.tsv file). To learn the basic concepts of working with the CLI, see this tutorial https://astrobiomike.github.io/unix/unix-intro by Michael Lee [33].
2. In this example, one of the aligners (*minimap*) produces an output that is directly fed to *samtools view* for conversion to binary format, and the output of *samtools view* is fed into *samtools sort.*
3. HPC environments are particularly valuable for computationally intensive tasks like genome assembly or population-scale

variant calling, as they provide access to multiple compute nodes, large memory capacity, and optimized job scheduling systems [34]. In simple terms, HPC is a set of independent computers (servers), interconnected so that the user can ask the scheduling system to execute a job (i.e., run a command) with some constraints (e.g., minimum memory, number of CPUs, maximum time), and the scheduler will put the task in a queue, waiting for one of the servers (called "nodes") of the HPC to be available to execute the task.

Acknowledgements

This work was funded by the Biotechnology and Biological Sciences Research Council (BBSRC) through an Institute Strategic Programme award to the QIB Programmes Gut Microbes and Health BB/R012490/1 and associated Core Capability Grant BB/CCG1860/1 and Food, Microbiome and Health (FMH) [BB/X011054/1] and associated Core Capability Grant BB/CCG2260/1.

References

1. Castrignanò T, Gioiosa S, Flati T et al (2020) ELIXIR-IT HPC@CINECA: high performance computing resources for the bioinformatics community. BMC Bioinformatics 21: 352. https://doi.org/10.1186/s12859-020-03565-8
2. Banimfreg BH (2023) A comprehensive review and conceptual framework for cloud computing adoption in bioinformatics. Healthcare Analytics 3:100190. https://doi.org/10.1016/j.health.2023.100190
3. Grzesik P, Augustyn DR, Wyciślik Ł, Mrozek D (2022) Serverless computing in omics data analysis and integration. Brief Bioinform 23: bbab349. https://doi.org/10.1093/bib/bbab349
4. Bai J, Jhaney I, Wells J (2019) Developing a reproducible microbiome data analysis pipeline using the Amazon web services cloud for a cancer research group: proof-of-concept study. JMIR Med Inform 7:e14667. https://doi.org/10.2196/14667
5. Brandies PA, Hogg CJ (2021) Ten simple rules for getting started with command-line bioinformatics. PLoS Comput Biol 17:e1008645. https://doi.org/10.1371/journal.pcbi.1008645
6. Mielczarek M, Czech B, Stańczyk J et al (2020) Extraordinary command line: basic data editing tools for biologists dealing with sequence data. TOBIOIJ 13:137–145. https://doi.org/10.2174/1875036202013010137
7. Raymond ES (2003) Art of UNIX programming, the, portable documents. Addison-Wesley Professional
8. The Bioconda Team, Grüning B, Dale R et al (2018) Bioconda: sustainable and comprehensive software distribution for the life sciences. Nat Methods 15:475–476. https://doi.org/10.1038/s41592-018-0046-7
9. da Veiga LF, Grüning BA, Alves Aflitos S et al (2017) BioContainers: an open-source and community-driven framework for software standardization. Bioinformatics 33:2580–2582. https://doi.org/10.1093/bioinformatics/btx192
10. The Galaxy Community, Afgan E, Nekrutenko A et al (2022) The galaxy platform for accessible, reproducible and collaborative biomedical analyses: 2022 update. Nucleic Acids Res 50: W345–W351. https://doi.org/10.1093/nar/gkac247
11. Barker M, Chue Hong NP, Katz DS et al (2022) Introducing the FAIR principles for research software. Sci Data 9:622. https://doi.org/10.1038/s41597-022-01710-x
12. Perez-Riverol Y, Gatto L, Wang R et al (2016) Ten simple rules for taking advantage of Git

and GitHub. PLoS Comput Biol 12: e1004947. https://doi.org/10.1371/journal.pcbi.1004947

13. Matthews TC, Bristow FR, Griffiths EJ et al (2018) The integrated rapid infectious disease analysis (IRIDA) platform. Bioinformatics
14. Di Tommaso P, Chatzou M, Floden EW et al (2017) Nextflow enables reproducible computational workflows. Nat Biotechnol 35:316–319. https://doi.org/10.1038/nbt.3820
15. Köster J, Rahmann S (2018) Snakemake-a scalable bioinformatics workflow engine. Bioinformatics 34:3600. https://doi.org/10.1093/bioinformatics/bty350
16. Ewels PA, Peltzer A, Fillinger S et al (2020) The nf-core framework for community-curated bioinformatics pipelines. Nat Biotechnol 38: 276–278. https://doi.org/10.1038/s41587-020-0439-x
17. Wikipedia (2025) Plain text. Wikipedia
18. Shen W, Sipos B, Zhao L (2024) SeqKit2: a Swiss army knife for sequence and alignment processing. iMeta 3:e191. https://doi.org/10.1002/imt2.191
19. Telatin A, Fariselli P, Birolo G (2021) SeqFu: a suite of utilities for the robust and reproducible manipulation of sequence files. Bioengineering 8:59. https://doi.org/10.3390/bioengineering8050059
20. Chen S, Zhou Y, Chen Y, Gu J (2018) Fastp: an ultra-fast all-in-one FASTQ preprocessor. Bioinformatics 34:i884–i890. https://doi.org/10.1093/bioinformatics/bty560
21. Quinlan AR, Hall IM (2010) BEDTools: a flexible suite of utilities for comparing genomic features. Bioinformatics 26:841–842. https://doi.org/10.1093/bioinformatics/btq033
22. Pertea G, Pertea M (2020) GFF utilities: GffRead and GffCompare. F1000Research, 9, ISCB-Comm. https://doi.org/10.12688/f1000research.23297.2
23. Dainat J, Hereñú D, Murray KD et al (2024) NBISweden/AGAT: AGAT-v1.4.1
24. Junier T, Zdobnov EM (2010) The Newick utilities: high-throughput phylogenetic tree processing in the Unix shell. Bioinformatics 26:1669–1670. https://doi.org/10.1093/bioinformatics/btq243
25. Li H, Handsaker B, Wysoker A et al (2009) The sequence alignment/map format and SAMtools. Bioinformatics 25:2078–2079. https://doi.org/10.1093/bioinformatics/btp352
26. McDonald D, Clemente JC, Kuczynski J et al (2012) The biological observation matrix (BIOM) format or: how I learned to stop worrying and love the ome-ome. Gigascience 1:7. https://doi.org/10.1186/2047-217X-1-7
27. McMurdie PJ, Holmes S (2013) Phyloseq: an R package for reproducible interactive analysis and graphics of microbiome census data. PLoS One 8:e61217. https://doi.org/10.1371/journal.pone.0061217
28. Oksanen J, Simpson GL, Blanchet FG, et al (2025) Vegan: community ecology package
29. Liu S, Wang Z, Zhu R et al (2021) Three differential expression analysis methods for RNA sequencing: limma, EdgeR, DESeq2. J Vis Exp. https://doi.org/10.3791/62528
30. Estaki M, Jiang L, Bokulich NA et al (2020) QIIME 2 enables comprehensive end-to-end analysis of diverse microbiome data and comparative studies with publicly available data. Curr Protoc Bioinformatics 70:e100. https://doi.org/10.1002/cpbi.100
31. Telatin A (2021) Qiime artifact eXtractor (qax): a fast and versatile tool to interact with Qiime2 archives. Biotech 10. https://doi.org/10.3390/biotech10010005
32. Ziemski M, Gehret L, Simard A et al (2025) MOSHPIT: accessible, reproducible metagenome data science on the QIIME 2 framework
33. Lee M (2019) Happy belly bioinformatics: an open-source resource dedicated to helping biologists utilize bioinformatics. JOSE 2:53. https://doi.org/10.21105/jose.00053
34. Suplatov D, Shegay M, Sharapova Y et al (2021) Co-designing HPC-systems by computing capabilities and management flexibility to accommodate bioinformatic workflows at different complexity levels. J Supercomput 77: 12382–12398. https://doi.org/10.1007/s11227-021-03691-x

Chapter 10

Bioinformatics Processing of 16S Datasets

Alise J. Ponsero, Anthony Duncan, Judit Talas, Katarzyna Sidorczuk, Wing Koon, Falk Hildebrand, and Andrea Telatin

Abstract

This chapter covers methods and best practices for analyzing 16S datasets, with a particular emphasis on the quality control steps, from the preprocessing of the raw sequencing reads to the removal of chimeric amplification products. It describes outputs from the analysis of raw reads that provide the foundation for downstream analyses and interpretation of the microbial community composition and diversity. It provides details of the tools to perform each step and how the framework Qiime2 and a Nextflow pipeline (nf-core/ampliseq) can improve the reproducibility and traceability of the analysis. Finally, it describes Lotus2, an efficient end-to-end pipeline developed at QIB, which allows the user to select the strategy to adopt for feature identification and taxonomic profiling.

Key words 16S rRNA, Bioinformatics, Quality control, Taxonomic profiling, Reproducibility, Traceability

1 Introduction

This chapter details the analysis of 16S amplicons, sequenced using Illumina platforms in paired-end mode. Illumina offers a high per-base quality and paired-end sequencing with the benefits of including sequencing amplicons longer than a single Illumina read (300 base pairs for MiSeq models and the currently available NextSeq 2000) and increased accuracy of the overlapping region [1].

A fundamental challenge in 16S analysis is distinguishing genuine biological sequences from artifacts introduced during PCR amplification and sequencing (noise). Two main approaches to addressing the noise are clustering sequences into Operational Taxonomic Units (OTUs) and denoising algorithms that attempt to recover true biological sequences, termed amplicon sequence variants (ASVs). While fundamentally different, these two methods produce comparable results, particularly after taxonomy assignment [2].

Simon R. Carding (ed.), *Best Practice in Microbiome Research*, Springer Protocols Handbooks,
https://doi.org/10.1007/978-1-0716-5009-7_10,

The OTU clustering approach, implemented in tools like USEARCH [3–5] and VSEARCH [6], groups sequences based on a similarity threshold (traditionally 97%). This reduces computational complexity and mitigates sequencing errors but may mask fine-scale biological variation. In contrast, denoising algorithms like DADA2 leverage error models learned from the sequencing data to infer the original biological sequences [7]. DADA2 considers quality scores and nucleotide transition probabilities to differentiate sequencing errors from true biological variation, potentially providing single-nucleotide resolution and better discrimination of closely related strains. Both USEARCH and DADA2 integrate algorithms for chimera removal [8].

After obtaining representative OTUs or ASVs, the next step is taxonomic assignment which relies on comprehensive reference databases such as SILVA, RDP, and Greengenes [9–11]. These databases enable taxonomy assignment through methods ranging from simple sequence similarity searches to more sophisticated phylogenetic placement approaches. The choice of reference database can significantly impact the taxonomic classifications, particularly for less well-characterized environmental communities.

The final output of a 16S analysis pipeline typically consists of four interconnected files:

1. The feature table (also known as an OTU or ASV table) representing the abundance of each feature across samples, structured as a matrix where rows correspond to features and columns to samples [12]
2. A taxonomy file that maps each feature to its taxonomic classification, often including confidence scores for each taxonomic level.
3. A phylogenetic tree of the features providing evolutionary context enabling phylogenetic diversity metrics.
4. A mapping file (or metadata file) containing sample information and experimental variables for downstream statistical analyses.

Together, these files enable comprehensive analyses of community composition, diversity measurements, and differential abundance testing.

Modern analysis frameworks like QIIME 2 integrate the steps described above into a reproducible workflow while maintaining detailed provenance tracking of all analytical steps [13, 14]. Similarly, the nf-core/ampliseq pipeline [15, 16], written in Nextflow, provides a standardized, containerized implementation that ensures reproducibility across computing environments (while incorporating multiple tools, allowing for high customization of the workflow by the user).

A dedicated end-to-end pipeline called Lotus2 [17], developed at the Quadram Institute, allows the user to analyze metabarcoding projects using configuration files to make analyses reproducible, offering performance gain and the flexibility of running the analysis on a local computer when a cluster is not needed.

2 Methods

2.1 Biases in Metabarcoding

While the overall goal of a metabarcoding experiment is to describe the taxonomic composition of a microbiota, a set of technical biases introduces distortions to the relative abundance of taxa in the final output. The systematic distortions begin at the sampling stage, where microhabitat heterogeneity and preservation conditions can substantially affect community representation, and continue through DNA extraction, where differential cell lysis efficiency and varying cell wall structures create taxonomic-specific extraction biases [18, 19].

The PCR amplification stage introduces perhaps the most substantial sources of bias [20], including primer binding preferences. These molecular biases are further compounded during library preparation and sequencing, where adapter ligation efficiency variations, PCR duplicates, and platform-specific sequencing errors contribute additional layers of technical variance. During PCR amplification, it is possible to generate chimeric products when incompletely extended primers from one DNA template anneal to and prime synthesis from a different but similar template during later PCR cycles. These chimeric sequences are essentially artificial hybrid molecules containing sequences from two different organisms spliced together.

The final bioinformatic analysis pipeline described in this chapter introduces its own set of computational considerations, from the choice of algorithm parameters to database completeness and taxonomic assignment methodologies [21, 22] (*see* **Note 1**).

2.2 Preprocessing Raw Reads

Understanding the structure of the 16S rRNA hypervariable region targeted guides the quality control of these sequences and ensures that the preprocessing step preserves the biological signal while removing low-quality sequences and artifacts.

Current sequencing machines produce a separate dataset per sample such that all the samples sequenced in the same sequencing run will be demultiplexed sorting reads according to a molecular barcode added (described in Chapter 7). If this is not the case, a demultiplexing step is required.

Primer removal is a critical step of 16S rRNA amplicon dataset preprocessing with incomplete primer sequence trimming having a significant impact on the downstream steps. The program Cutadapt [23] can be used for this task (*see* **Note 2**). Importantly, reads without primers should be excluded from the downstream analysis as they likely arise from technical artifacts. Finally, primer dimers can be excluded by filtering the read length after trimming. Note, it is good practice to collate and report the percentage of read pairs where primers are detected, to investigate eventual outliers with significant deviation from the average. When processing the reads with DADA2, primer removal is particularly important (see below).

After primer and adapter removal/trimming, Fastp [24] can be used for quality filtering and removing reads with very-low-quality scores by setting a minimum length (*see* **Note 3**).

After primer trimming and quality filtering, reads from paired-end sequencing data are merged to cover the full amplified region. *VSEARCH* or *DADA2* (its mergePairs function) merges the paired-end reads and chooses appropriate parameters for the expected overlap region. The merging parameters should be determined by the amplification hypervariable region length and the sequencing read length. While shorter target region requires a longer sequence overlap for the pairs, longer regions such as the V3–V4 regions need more lenient overlap parameters. The proper implementation of this merging step provides an additional quality control, as inconsistent overlap regions will be discarded (*see* **Note 4**).

Community pipelines, such as the nf-core/ampliseq pipeline [16], can be used to ensure consistent preprocessing across projects and allow users to easily generate a comprehensive documentation of the preprocessing parameters applied.

2.3 Identifying Representative Sequences

Representative sequences in microbial metabarcoding serve as molecular identifiers that characterize the diversity of microbial communities which typically fall into two categories: operational taxonomic units (OTUs) that cluster similar sequences under a representative sequence and amplicon sequence variants (ASVs) that emerge from denoising without clustering (*see* **Note 5**).

When analyzing Illumina metabarcoding datasets, denoising is most often used especially since the introduction of the DADA2 package. Denoised sequences, referred to as ASVs (or exact sequence variants) [25], increase the power of resolution of the analysis and, if needed, allow for grouping results at the same taxonomic level.

The feature table (also referred to as OTU table, or ASV table) records the abundance of each representative sequence across samples, where rows represent the sequences (ASV or OTUs) and columns represent samples, with each cell containing the count of that sequence in that sample. This table is generated during the denoising with DADA2, and alternatively, programs like VSEARCH (using the "usearch_global" function) can generate one starting from a feature file (FASTA) and the filtered reads for each sample (FASTQ).

2.4 Taxonomic Assignment

Taxonomic classification of ASVs involves matching sequences against curated reference databases using specialized algorithms. This process assigns taxonomic ranks hierarchically, from kingdom to species (typically with decreasing confidence at more specific ranks). Several reference databases serve this purpose [26], each with distinct strengths: RDP [9], SILVA [10], and Greengenes [11] focus on prokaryotic ribosomal RNA genes, while UNITE [27] specializes in fungal sequences. SILVA is widely adopted due to its comprehensive curation, regular updates, and extensive coverage of environmental taxa, all supported by quality control and alignment with standard taxonomic frameworks. The *ensembleTax* package offers a solution to leverage multiple databases and classification methods simultaneously, combining their individual strengths through two key functionalities [28].

Common classification tools include the RDP Classifier, which uses a Bayesian approach, SINTAX with its *k*-mer similarity method, and DECIPHER's tree-based classification [29–31]. A commonly used approach is the q2-feature-classifier which harbors a machine learning framework with flexible k-mer lengths (up to 32-mers, unlike the fixed 8-mers of RDP's Naïve Bayesian). When evaluated against mock communities, this approach demonstrates enhanced precision particularly at lower taxonomic ranks, while maintaining high recall rates comparable to established methods [30]. Each tool provides confidence scores for taxonomic assignments, helping researchers evaluate classification reliability (*see* **Note 6**).

2.5 Evaluating the Phylogenetic Distance of ASVs

The phylogenetic placement of ASVs is an optional but important step in the analysis. A tree of features can be used for understanding the evolutionary relationships between microbial taxa and enables the calculation of phylogeny-aware beta diversity metrics like UniFrac. UniFrac incorporates both presence/absence and abundance data while accounting for the evolutionary distances between observed organisms, providing deeper insights into community differences than traditional distance metrics (Chapter 9).

The first step is to perform a multiple sequence alignment, using a tool such as MAFFT or SINA (which accounts for secondary structure conservation, improving alignment quality compared to general-purpose tools). The aligned sequences then require masking to remove highly variable regions using tools like trimAl or SINA's built-in masking.

Tree construction should employ maximum likelihood methods through IQ-TREE2, which provides more accurate evolutionary reconstructions than distance-based methods. IQ-TREE2 provides a "ModelFinder Plus" option to automatically select the best evolutionary model and perform 1000 ultrafast bootstraps to assess branch support. When computational resources are limited,

FastTree is an alternative tool that provides a reasonable compromise between speed and accuracy reliability (*see* **Note 7**).

2.6 Performing Multiple Steps Using DADA

While conceptually separated, DADA2 is a single tool that can be used for multiple steps.

2.6.1 Quality Control and Filtering of Raw Reads

Use `fastqFilter()` or `filterAndTrim()` to process raw fastq files to (1) remove low-quality sequence ends, (2) filter out reads with too many expected errors, and (3) remove PhiX contamination. Primers can be trimmed with the `removePrimers()` function.

2.6.2 Read Merging (for Paired-End Data)

Use `mergePairs()` to combine the filtered forward and reverse reads. This step aligns the reads (the forward pair and the reverse-complemented reverse pair) on overlapping regions and creates consensus sequences from the merged pairs.

2.6.3 Sequence Denoising

The key step performed by DADA is the `dada()` function, which performs error correction and sequence inference. First, learn the error rates from the data using `learnErrors()`. This must be performed on the biological sequences, after removal of adapters and primers, and is run-dependent (i.e., if samples were sequenced in multiple sequencing runs), applies the error model to detect and correct sequence errors, and infers the real biological sequences present in the samples.

2.6.4 Generating the Count Table

Use `makeSequenceTable()` to create an abundance matrix (the *feature table*). The output table has rows representing samples and columns representing unique sequences (ASVs), and each value represents the number of times each sequence appears in each sample.

2.6.5 Chimera Removal

Use `removeBimeraDenovo()`to detect and remove chimeric sequences by first identifying sequences that appear to be artificial combinations of other sequences and then removing them from the dataset.

2.6.6 Taxonomy Assignment

Use `assignTaxonomy()`to match sequences against a reference database.

2.7 Integrating Datasets for Downstream Analyses

As outlined in Chapter 10, the key outputs of a metabarcoding analysis are as follows:

1. *A mapping file*—a tabular file with the attributes of each sample. For human studies, we can have attributes of the sample donor (such as age, BMI, pathologies...) and attributes of the sample itself (collection date, DNA extraction date...)

2. *A feature table*—a matrix reporting all the identified sequences (ASVs) as rows, and the time each feature was found in each sample (columns).
3. *The taxonomy*—a tabular file assigning the predicted taxonomy for each feature.
4. *The feature tree*—a Newick tree of the representative sequences.

These files must be coherent: The sample identifier must be consistent between metadata and the feature table, and the names of the features must be consistent across the feature table, tree, and taxonomy.

Phyloseq [32] is a specialized R package designed for comprehensive microbiome data analysis, serving as an integrated framework for analyzing, visualizing, and exploring complex microbial community datasets. A PhyloSeq object is created starting from the components noted above:

```
physeq <- phyloseq(FEATURE_TABLE, TAX_TABLE, MAPPING_TABLE,
TREE)
```

The Phyloseq package implements a comprehensive data model that combines abundance measurements, taxonomic assignments, phylogenetic relationships, and sample metadata. Its core functionality centers on efficient data manipulation, statistical analysis, and visualization of complex microbial community datasets.

2.8 Qiime2 Created Frameworks with FAIR Artifacts

A metabarcoding analysis requires several steps performed using different tools. When an output file is saved (for example, a feature counts table), the researcher cannot tell which methods were used upstream to generate it and must rely on their ability to track the work done.

Qiime2 wraps all the steps of a metabarcoding analysis in a coherent framework, where metadata is embedded in each input and output file (called artifacts in the Qiime2 nomenclature, and recognizable by their ".qza" or ".qzv" extensions). Using Qiime2, each file recapitulates all the steps used to generate it [14] (see Chapter 9).

A Qiime2 analysis starts by importing datasets into an artifact, for example, a set of FASTQ reads: this can be performed with the command:

```
qiime tools import \
 --type 'SampleData[PairedEndSequencesWithQuality]' \
 --input-path manifest.txt \
 --output-path paired-end.qza \
 --input-format PairedEndFastqManifestPhred33V2
```

where the "manifest.txt" file is a CSV (comma-separated values) with sample identifiers and path to the FASTQ files of each sample, and "paired-end.qza" is a single artifact containing all the reads, ready to be used for downstream analyses.

Qiime2 provides a coherent interface to run all the programs used in the metabarcoding analysis. For example, if running DADA2 as embedded in Qiime2, then run the command:

```
qiime dada2 denoise-paired \
 --i-demultiplexed-seqs paired-end.qza \
 --p-trim-left-f 0 \
 --p-trim-left-r 0 \
 --p-trunc-len-f 250 \
 --p-trunc-len-r 250 \
 --p-n-threads 0 \
 --o-representative-sequences rep-seqs.qza \
 --o-table table.qza \
 --o-denoising-stats stats.qza
```

The advantage of using Qiime2 is that it ensures only compatible file types are used (for example, a FASTA file cannot be used as input).

The Qiime2 website (https://use.qiime2.org/) contains extensive documentation and a set of tutorials guiding the user in performing the analysis from the raw reads to the exploration of the diversity of the samples and statistical analysis.

2.9 Nextflow-Based Reproducible and Scalable Analysis

The **nf-core/ampliseq** pipeline represents a significant advancement in the standardization and reproducibility of 16S rRNA gene amplicon sequence analysis (and other universal marker genes). This computational framework integrates state-of-the-art bioinformatics tools while adhering to the FAIR principles (Chapter 9), making it an invaluable resource for microbiome research.

The pipeline is built upon Qiime2 with DADA2 as its core. A key strength of nf-core/ampliseq lies in its simplified management of software dependencies, requiring only Nextflow, Java, and a container executor such as Apptainer or Docker.

The modules embedded in the pipeline are shown in Fig. 1, separated into the main steps:

1. Quality filtering, detection of representative sequences (Infer ASVs).
2. Taxonomic assignment and filtering.
3. Generation of a set of visualizations and final report.

The key steps enabled by default (highlighted in green) are the steps presented in this chapter, in particular primer removal with

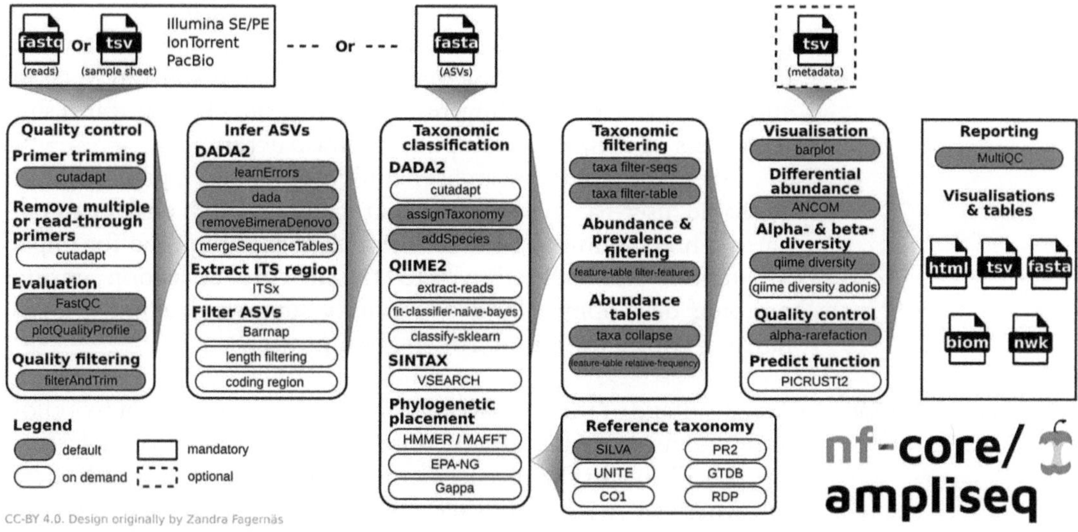

Fig. 1 Schematic overview of the step (and tools) implemented in nf-core/ampliseq. The modules enabled by default are highlighted in green

Cutadapt, denoising, and chimera removal and taxonomic assignment (using *DADA2*).

The analytical workflow begins with raw sequence data processing, where the only essential user inputs are the folder containing raw sequencing data and primer sequences. The pipeline then processes these data through multiple analytical stages, including quality control assessment, denoising procedures, and taxonomic classification. The framework automatically handles computational resource allocation, making efficient use of high-performance computing clusters or cloud computing infrastructure through containerized applications.

A distinguishing feature of nf-core/ampliseq is its comprehensive output generation system. The pipeline produces quality metrics for raw read data and denoising processes, alongside interactive visualizations of community composition through bar plots. The organized output folder includes alpha and beta diversity measurements, ordination plots, and differential abundance analyses with associated statistical tests. The pipeline incorporates filtering capabilities for removing unwanted sequences, with default settings targeting common contaminants such as mitochondrial and chloroplast sequences. Users can modify these filters based on taxa, prevalence, or count thresholds, providing flexibility in data cleaning approaches. Furthermore, the integration of metadata enables group comparisons and statistical analyses, essential for addressing complex ecological questions.

The reproducibility of analyses is ensured through the pipeline's container-based approach, where all software dependencies are bundled and automatically deployed during analysis. This design choice addresses a persistent challenge in bioinformatics—the reproduction of results across different computational environments. Each pipeline release maintains consistent software versions and dependencies, enabling researchers to obtain identical results regardless of their local computing infrastructure.

For microbiome research, the nf-core/ampliseq pipeline offers a reliable, user-friendly solution that maintains high analytical standards while minimizing the technical expertise required for implementation. Its integration into the nf-core collection ensures ongoing development and community support, making it a sustainable choice for long-term research projects.

2.9.1 Configuration File

An advantage of the Nextflow pipeline is the separation of workflow logic and its configuration. A configuration file can be considered as a protocol to run the nf-core/ampliseq pipeline on a specific dataset, where we specify input files, parameters, databases, and technical details (container engine to use, where to run the tasks, ...).

2.10 Lotus2 Provides a Fast and Customizable Pipeline

LotuS2 (less OTU scripts 2, available at https://lotus2.earlham.ac.uk/) is designed to be accessible and versatile [17]. Using a single command, users can input either long or short sequencing reads and compute the abundance matrices, taxonomic assignments, and phylogenies in a variety of formats, including PhyloSeq objects. LotuS2 is highly customizable to specific experimental conditions, offering 73 flags and 32 options within configuration files to enable more experienced users to customize data processing. Default configurations are made available for various scenarios, making the pipeline accessible to those with little bioinformatic experience or who have limited time to fine-tune analysis.

The most basic use of LotuS2 requires the input path (`-i`), output directory (`-o`), and path to the mapping file (`-m`) which contains information on sample identifiers, demultiplexing barcodes, or file paths to demultiplexed samples. In Linux bash, this could look like the following:

```
lotus2 -i Example/ -m Example/miSeqMap.sm.txt -o OutputDirectory
```

Additional flags can be grouped into four categories: (1) miscellaneous, (2) workflow related, (3) taxonomy, and (4) clustering options. Clustering options define whether OTUs or ASVs will be created via five different algorithms, as well as handling chimera checks and dereplication settings. Taxonomy options are used to fine tune the taxonomic assignments, offering support for different amplicon types, seven algorithms to assign a taxonomy to OTUs/ASVs against five installed databases, and the option to provide custom reference databases. Miscellaneous options allow for the utilization of temporary directories and applying various configuration files where desired. Workflow options control the general path an analysis takes, including primer, chimera, cross-talk, and host-contamination checks, or skipping most of the pipeline to conduct taxonomic classification only.

Upon installation, LotuS2 has a series of configuration files that are written for different sequencing platforms (so-called sdm configurations), which are then used to control read filtering, with parameters such as minimal read quality, accumulated errors, whether to reorient reads, scan, and remove remaining sequencing and amplification primers, among others. Based on the output, users could alter these files, which will be discussed later (https://github.com/hildebra/sdm).

LotuS2 is on average 29 times faster than the most used pipelines, enabling users to easily fine-tune and re-run their analyses without the need for intensive computational power. Upon completion, the output log files are indicative of the parameters that can be optimized. For example, *demulit.log* contains information on the number of reads rejected and why. Based on the reasoning, if suitable, thresholds can be adjusted within the sdm config files, maximizing the number of reads that can pass through filtering and retain good quality data.

3 Notes

1. A take-home message is that, if applying the same protocol throughout the study, meaningful across-samples comparisons can still be performed. Comparing the relative abundances of taxa within samples is not possible; in other words, if one taxon has double the relative abundance in the bioinformatics results, it cannot be assumed it was the case in the original community.
2. With paired-end datasets will simultaneously remove both adapters from each pair, and it supports IUPAC degenerate bases. The command is like.

```
cutadapt \
-a {FWDPRIMER}...{RC_REVPRIMER} -A {REVPRIMER}...{RC_FWDPRI-
MER} \
--discard-untrimmed \
-o out_R1.fastq.gz -p out_R2.fastq.gz \
in.1.fastq.gz in.2.fastq.gz
```

3. Characteristics of amplicon datasets should be taken into consideration during the preprocessing stage. Too aggressive trimming of low-quality bases can prevent the read merging step and affect all downstream processing. A relatively lenient quality filtering and relying on the denoising step in the

processing pipeline for error corrections is therefore recommended.

4. It is recommended that the number of reads discarded at each filtering step be tracked, as a reduction in reads at a specific step is indicative of a technical issue or a potential need for parameter tuning. Importantly, all preprocessing steps should be clearly documented to ensure reproducibility. Indeed, different preprocessing approaches can lead to significantly different results.
5. These sequences do not map one-to-one with bacterial species: A single V3-V4 region of the 16S gene can be identical across multiple species, while some taxa possess multiple, slightly divergent copies of the rDNA operon (e.g., >1.5% nucleotide difference in Vibrionaceae).
6. Several factors impact classification accuracy: Short amplicon sequences may lack sufficient information for species-level assignment, reference databases often miss environmental taxa, and inconsistent nomenclature across databases can complicate comparisons between studies. Whichever method is used, it is crucial to report confidence thresholds and handle unclassified taxa appropriately in downstream analyses, and it's advisable to consider metabarcoding classifications up to the genus level.
7. Best practices include filtering ASVs to remove sequences appearing in fewer than 0.1% of samples before tree construction, including reference sequences from well-characterized taxa as guides for phylogenetic placement, validating tree topology through bootstrap support values. The output file is a tree in Newick format.

Acknowledgements

This work was funded by the Biotechnology and Biological Sciences Research Council (BBSRC) through an Institute Strategic Programme award to the QIB Programmes Gut Microbes and Health BB/R012490/1 and associated Core Capability Grant BB/CCG1860/1 and Food, Microbiome and Health [BB/X011054/1] and associated Core Capability Grant BB/CCG2260/1.

References

1. Kozich JJ, Westcott SL, Baxter NT et al (2013) Development of a dual-index sequencing strategy and curation pipeline for analyzing amplicon sequence data on the MiSeq Illumina sequencing platform. Appl Environ Microbiol 79:5112–5120. https://doi.org/10.1128/AEM.01043-13

2. Glassman SI, Martiny JBH (2018) Broadscale ecological patterns are robust to use of exact sequence variants versus operational taxonomic units. mSphere 3:e00148-18. https://doi.org/10.1128/mSphere.00148-18
3. Edgar RC (2010) Search and clustering orders of magnitude faster than BLAST. Bioinformatics 26:2460–2461. https://doi.org/10.1093/bioinformatics/btq461
4. Edgar RC (2016) UNOISE2: improved error-correction for Illumina 16S and ITS amplicon sequencing. 081257
5. Edgar RC (2013) UPARSE: highly accurate OTU sequences from microbial amplicon reads. Nat Methods 10:996–998. https://doi.org/10.1038/nmeth.2604
6. Rognes T, Flouri T, Nichols B et al (2016) VSEARCH: a versatile open source tool for metagenomics. PeerJ 4:e2584. https://doi.org/10.7717/peerj.2584
7. Callahan BJ, McMurdie PJ, Rosen MJ et al (2016) DADA2: high-resolution sample inference from Illumina amplicon data. Nat Methods 13:581–583. https://doi.org/10.1038/nmeth.3869
8. Edgar RC, Haas BJ, Clemente JC et al (2011) UCHIME improves sensitivity and speed of chimera detection. Bioinformatics 27:2194–2200. https://doi.org/10.1093/bioinformatics/btr381
9. Cole JR, Wang Q, Fish JA et al (2014) Ribosomal database project: data and tools for high throughput rRNA analysis. Nucleic Acids Res 42:D633–D642. https://doi.org/10.1093/nar/gkt1244
10. Quast C, Pruesse E, Yilmaz P et al (2013) The SILVA ribosomal RNA gene database project: improved data processing and web-based tools. Nucleic Acids Res 41:D590–D596. https://doi.org/10.1093/nar/gks1219
11. McDonald D, Jiang Y, Balaban M et al (2024) Greengenes2 unifies microbial data in a single reference tree. Nat Biotechnol 42:715–718. https://doi.org/10.1038/s41587-023-01845-1
12. Regueira-Iglesias A, Balsa-Castro C, Blanco-Pintos T, Tomás I (2023) Critical review of 16S rRNA gene sequencing workflow in microbiome studies: from primer selection to advanced data analysis. Mol Oral Microbiol 38:347–399
13. Bolyen E, Rideout JR, Dillon MR et al (2019) Reproducible, interactive, scalable and extensible microbiome data science using QIIME 2. Nat Biotechnol 37:852–857. https://doi.org/10.1038/s41587-019-0209-9
14. Estaki M, Jiang L, Bokulich NA et al (2020) QIIME 2 enables comprehensive end-to-end analysis of diverse microbiome data and comparative studies with publicly available data. Curr Protoc Bioinformatics 70:e100. https://doi.org/10.1002/cpbi.100
15. Ewels PA, Peltzer A, Fillinger S et al (2020) The nf-core framework for community-curated bioinformatics pipelines. Nat Biotechnol 38:276–278. https://doi.org/10.1038/s41587-020-0439-x
16. Straub D, Blackwell N, Langarica-Fuentes A et al (2020) Interpretations of environmental microbial community studies are biased by the selected 16S rRNA (gene) amplicon sequencing pipeline. Front Microbiol 11:550420. https://doi.org/10.3389/fmicb.2020.550420
17. Özkurt E, Fritscher J, Soranzo N et al (2022) LotuS2: an ultrafast and highly accurate tool for amplicon sequencing analysis. Microbiome 10:176. https://doi.org/10.1186/s40168-022-01365-1
18. Brooks JP, Edwards DJ et al (2015) The truth about metagenomics: quantifying and counteracting bias in 16S rRNA studies. BMC Microbiol 15:66. https://doi.org/10.1186/s12866-015-0351-6
19. Pollock J, Glendinning L, Wisedchanwet T, Watson M (2018) The madness of microbiome: attempting to find consensus "best practice" for 16S microbiome studies. Appl Environ Microbiol 84:e02627–e02617. https://doi.org/10.1128/AEM.02627-17
20. Nichols RV, Vollmers C, Newsom LA et al (2018) Minimizing polymerase biases in metabarcoding. Mol Ecol Resour 18:927–939. https://doi.org/10.1111/1755-0998.12895
21. O'Rourke DR, Bokulich NA, Jusino MA et al (2020) A total crapshoot? Evaluating bioinformatic decisions in animal diet metabarcoding analyses. Ecol Evol 10:9721–9739. https://doi.org/10.1002/ece3.6594
22. Tapolczai K, Keck F, Bouchez A et al (2019) Diatom DNA Metabarcoding for biomonitoring: strategies to avoid major taxonomical and Bioinformatical biases limiting molecular indices capacities. Front Ecol Evol 7:409. https://doi.org/10.3389/fevo.2019.00409
23. Martin M (2011) Cutadapt removes adapter sequences from high-throughput sequencing reads. EMBnet j 17:10. https://doi.org/10.14806/ej.17.1.200
24. Chen S, Zhou Y, Chen Y, Gu J (2018) Fastp: an ultra-fast all-in-one FASTQ preprocessor. Bioinformatics 34:i884–i890. https://doi.org/10.1093/bioinformatics/bty560

25. Callahan BJ, McMurdie PJ, Holmes SP (2017) Exact sequence variants should replace operational taxonomic units in marker-gene data analysis. ISME J 11:2639–2643. https://doi.org/10.1038/ismej.2017.119

26. Santamaria M, Fosso B, Consiglio A et al (2012) Reference databases for taxonomic assignment in metagenomics. Brief Bioinform 13:682–695. https://doi.org/10.1093/bib/bbs036

27. Abarenkov K, Henrik Nilsson R, Larsson K-H et al (2010) The UNITE database for molecular identification of fungi – recent updates and future perspectives. New Phytol 186:281–285. https://doi.org/10.1111/j.1469-8137.2009.03160.x

28. Catlett D, Son K, Liang C (2021) ensembleTax: an R package for determinations of ensemble taxonomic assignments of phylogenetically-informative marker gene sequences. PeerJ 9:e11865. https://doi.org/10.7717/peerj.11865

29. Wright ES (2016) Using DECIPHER v2.0 to analyze big biological sequence data in R. R J 8: 352. https://doi.org/10.32614/RJ-2016-025

30. Bokulich NA, Kaehler BD, Rideout JR et al (2018) Optimizing taxonomic classification of marker-gene amplicon sequences with QIIME 2's q2-feature-classifier plugin. Microbiome 6: 90. https://doi.org/10.1186/s40168-018-0470-z

31. Wang Q, Garrity GM, Tiedje JM, Cole JR (2007) Naïve Bayesian classifier for rapid assignment of rRNA sequences into the new bacterial taxonomy. Appl Environ Microbiol 73:5261–5267. https://doi.org/10.1128/AEM.00062-07

32. McMurdie PJ, Holmes S (2013) Phyloseq: an R package for reproducible interactive analysis and graphics of microbiome census data. PLoS One 8:e61217. https://doi.org/10.1371/journal.pone.0061217

Chapter 11

Bioinformatics Processing of Whole Metagenome Shotgun Datasets

Anthony Duncan, Wing Koon, Katarzyna Sidorczuk, Alise J. Ponsero, Sumeet K. Tiwari, Falk Hildebrand, and Andrea Telatin

Abstract

This chapter presents current best practices for the bioinformatic analysis of Whole Metagenome Sequencing (WMS) datasets discussing key methodological challenges. Read-based analysis of WGS enables the taxonomic classification and functional profiling through comparison of the sequencing reads against reference databases with assembly-based analysis allowing recovery of metagenome-assembled genomes (MAGs). Together, both approaches offer a complementary insight into the microbiome and require users to navigate the vast number of tools and methodologies published. Practical recommendations for tool selection and parameter optimization, considering computational requirements and biological accuracy, are included throughout.

Key words Metagenomics, Taxonomic profiling, Metagenome assembled genomes, Reproducibility

1 Introduction

WMS provides a microbiome's taxonomic composition and its functional potential, enabling identification of novel microorganisms and the description of previously unknown metabolic pathways in various ecosystems. Critically, the bioinformatics analysis of WMS datasets represents a significant computational challenge, due to their size and complexity, typically encompassing billions of short DNA sequences derived from thousands of different organisms. Consequently, the analysis of WMS datasets requires scalable bioinformatics tools able to process massive datasets in a reasonable timeframe. This chapter focuses on two main strategies to analyze WMS datasets: read-based analysis and assembly-based analysis. Read-based analyses rely on a reference database to perform taxonomic classification and functional annotation of sequencing reads. However, this approach is limited by the completeness of the reference database, as many organisms lack close relatives in it. This leads to a proportion of sequencing reads remaining

Simon R. Carding (ed.), *Best Practice in Microbiome Research*, Springer Protocols Handbooks, https://doi.org/10.1007/978-1-0716-5009-7_11,

unclassified or misclassified. On the other hand, assembly-based methods focus on reconstructing longer genomic sequences from the short reads, ultimately enabling the recovery of metagenome-assembled genomes (MAGs). Ideally, this approach facilitates the characterization of novel organisms and metabolic pathways, but it can be hindered by uneven sequence coverage, strain-level variation, and the presence of repetitive elements. This process is summarized in Fig. 1.

This chapter provides a comprehensive overview of the best practices for processing WMS datasets and includes a custom pipeline, mg-tk, that gives users greater flexibility in their analysis of WMS datasets.

2 Methods

2.1 Removing Host Reads

The presence of host DNA not only raises privacy concerns when working with human samples but can also impact downstream analyses. Importantly, the standard approach of aligning reads to the human reference genome is often insufficient on its own, and the remaining human reads can dramatically affect the conclusions of a study, in particular when processing low microbial biomass samples for which these residual host reads can lead to false conclusions [1].

Efficiently removing host reads is a significant computational challenge: First, the reference genome may be incomplete or altogether not available when dealing with non-human hosts. Additionally, sequencing errors can generate sequences that fail to perfectly match the reference genome, thereby complicating identification. Finally, large-scale genomic differences between samples and reference genomes, particularly structural variations, can prevent proper read mapping.

Using specialized decontamination tools, such as HOSTILE [2], which align reads against recent host reference genomes is recommended. These tools also provide users with the flexibility to build a custom host reference dataset. They maximize host read identification while limiting false-positive exclusions. When possible, use masked versions of the host genomes against known bacterial and viral sequences to avoid the false exclusion of microbial reads that share similarity with regions in the host genome. For any analysis in which the remaining host contamination could have a significant impact, ensuring effective host read removal requires a second validation approach (*see* **Note 1**).

Residual host content should be quantified using independent methods, such as k-mer-based tools or alignment to alternative references. Whenever possible, such as during read-based taxonomic profiling, the inclusion of the host genome in the reference database can help reduce false-positive assignments.

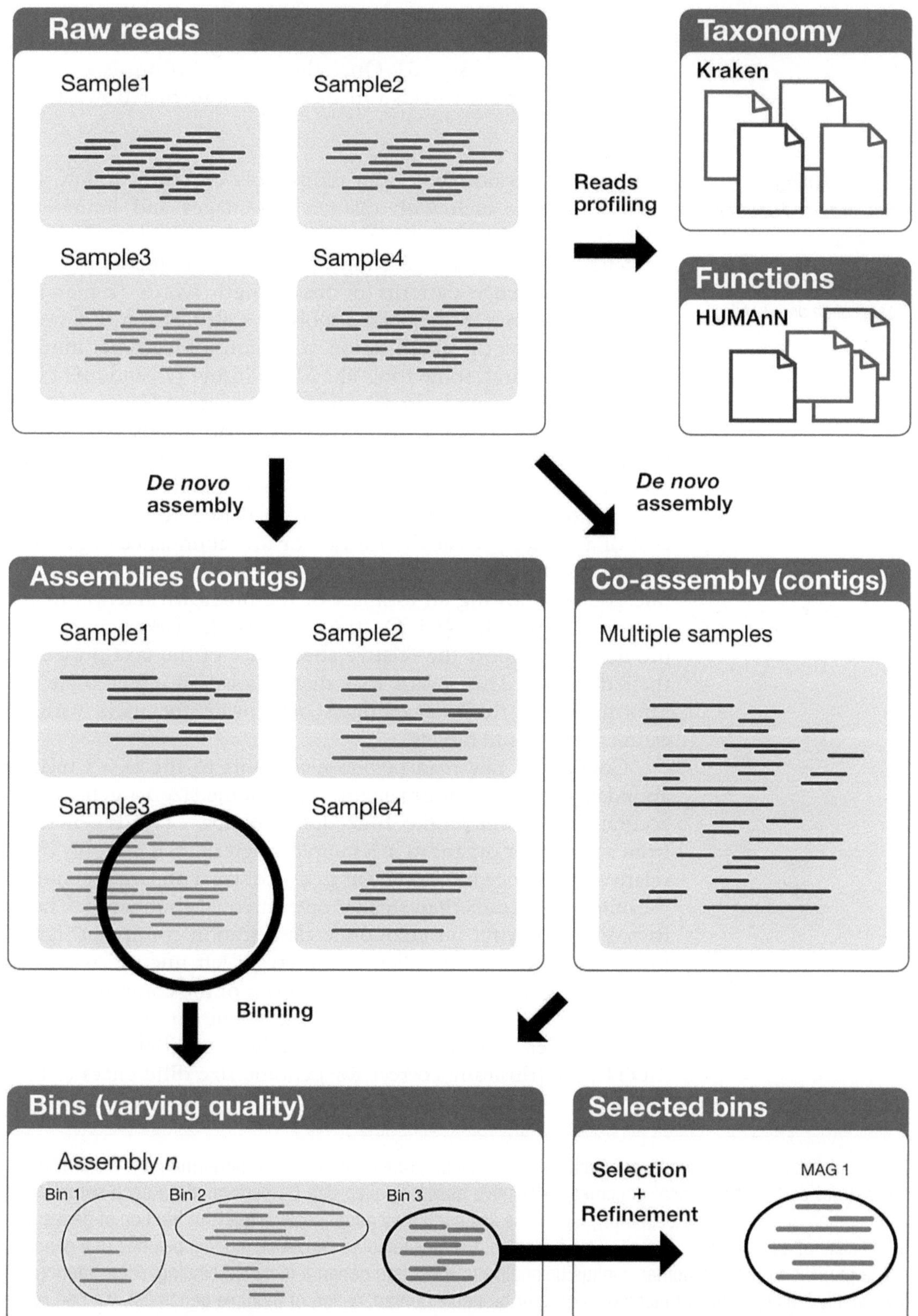

Fig. 1 Raw reads from microbiome samples can be processed independently from each other with read-based profilers (to define the taxonomic composition if using taxonomic profilers, such as Kraken or MetaPhlAn, or

After host-reads removal, it is important to trim the primers/adapters as well as the low-quality reads or bases, which can be achieved by using fastp [3]. Once these quality controls are completed, the processed reads can be used for further downstream analysis.

2.2 Assigning Taxonomy or Function to Reads

2.2.1 Taxonomic Classification and Profiling

Read-based taxonomic classification tools currently follow two main strategies, each with distinct advantages and limitations. First, the "*k*-mer based approach" was implemented by tools like Kraken2 [4], Centrifuge [5], and CLARK [6] and relies on matching exact sequence patterns of fixed-length words (k-mers) to reference genomes. K-mer-based tools are scalable to large datasets, typically allowing the classification of millions of reads per minute. On the other hand, some tools like MetaPhlAn4 [7] and mOTUs2 [8] use a marker gene approach which identifies reads in the dataset carrying specific genomic regions evolutionarily conserved within microbial taxa. Typically, these tools provide a more accurate read classification than k-mer-based approaches that are more prone to false-positive matches. However, this increased accuracy is at the cost of a reduced sensitivity for rare or low-abundance organisms. K-mer-based tools explicitly report the proportion of reads left unclassified, allowing an estimate of the unknown fraction of the microbial population [9]. On the other hand, marker-gene-based profilers only report the relative abundance of the taxa present in their database. This means that these tools will either omit this information in the metagenomes or provide the users with an estimate of the unknowns.

Converting raw read taxonomic counts to the taxa's relative abundances requires normalizing for genome size to generate an accurate taxonomic profile. Indeed, the number of reads generated from a particular organism in a sample is determined not only by its relative abundance but also by its genome size with larger genomes meaning more reads than smaller ones at equal abundance. Therefore, simply relying on taxonomic classification count can significantly bias the taxa abundance estimates if left uncorrected when integrating counts from different fractions of the community, such as the viral and bacterial fractions, and when genome size differences between organisms are particularly large. MetaPhlAn4 and mOTUs3 intrinsically correct for genome size differences as they

Fig. 1 (continued) the functional composition if using protein markers as performed by HUMAnN). To reconstruct part of the original genomes, a de novo assembly step can be performed on each sample or a combination of samples (co-assembly) when it is assumed they might share a relevant fraction of genes, like in the case of a longitudinal sampling study design. Each assembled sequence (contig) belongs to a genome, and a process to group together contigs belonging to the same genome is called binning. After filtering out low-quality bins (because of incompleteness or probable contamination of multiple genomes), we can refine bins to produce metagenome-assembled genomes (MAGs)

identify and quantify specific marker single-copy genes across genomes, providing a natural normalization. Certain k-mer-based tools such as Centrifuge integrate a normalization step; however, for tools like Kraken2, additional normalization steps are necessary. The bracken tool [10] can be applied to Kraken2 results to estimate species abundances by considering both genome length and expected coverage.

The choice of reference database can also impact taxonomic profiling results [11]. Two main types of databases can be leveraged by profiling tools. First, genome databases like NCBI RefSeq, which contain only curated and well-characterized genomes, provide high-confidence assignments but may miss novel organisms and are typically biased toward human-associated and clinically relevant microbes. Second, resources like GTDB-Tk [12] and MetaPhlAn4's database contain both isolated genomes and high-quality MAGs generated from metagenomic datasets that aim to cover a broader microbial diversity, but at the cost of reduced certainty in taxonomic assignments.

Given the vast amount of available taxonomic profiling tools and databases, interested users should consult comprehensive benchmarking studies and tool comparisons before selecting their computational method. For example, the Critical Assessment of Metagenome Interpretation (CAMI) challenges and OPAL [13, 14] provide systematic evaluations of read-based taxonomic profiling tools across diverse datasets. Additionally, standardized computational workflows such as nf-core/taxprofiler allow users to simultaneously use multiple taxonomic profiler tools and databases and implement best practices for each method. This pipeline generates results in consistent formats, enabling the direct comparison between different approaches and facilitating the validation of taxonomic assignments through consensus methods.

2.2.2 Functional Profiling

In WMS datasets, functional profiling allows investigations of its metabolic potential and biological capabilities present in global microbial communities. General-purpose functional profilers such as HUMANn3 [15] allow characterization of the overall community function. These tools typically rely on comprehensive protein databases to identify and quantify common metabolic pathways and protein families. This is particularly valuable if the goal is to characterize major well-known metabolic pathways and core microbial functions from the metagenome. However, when investigating specific functional categories, specialized profilers often provide superior sensitivity and annotation accuracy. For example, dbCAN3 [16] allows profiling of carbohydrate-active enzymes and leverages a carefully curated reference database along with specialized models to perform this annotation. This more focused approach can detect subtle variations in protein families and often provides more detailed functional annotations than general-purpose tools. Performing functional read profiling of a metagenomic dataset, particularly when using a general-purpose profiler, requires intensive computational requirements. Therefore, when planning the feasibility of their functional analysis, users need to consider the availability of computational resources (*see* **Note 2**).

Based on a read-based profiling approach, it is difficult to distinguish the origin(s) of a particular pathway. Notably, read-based functional annotation can be complemented by assembly-based methods. Indeed, the creation of a custom gene catalog from the WMS dataset can be of particular interest when working in understudied environments or when focusing on specific pathways of interest. The generation and use of custom gene catalogs is discussed in Subheading 2.4.

2.3 Generating High-Quality MAGs

While read-based approaches allow for profiling the taxonomic and functional potential of a community, they are typically limited to the detection of known organisms using reference databases. Complementary to these approaches, assembly-based methods reconstruct longer genomic sequences through assembly and binning.

2.3.1 Assembly

The first step in reconstructing MAGs from metagenomic reads is the assembly of short reads into contigs which is computationally intensive and complex due to the presence of many different organisms in the metagenome. Short-read assemblers like MEGAHIT [17] or metaSPADES [18] were developed to address these challenges and are scalable to large metagenomic datasets (*see* **Note 3**).

A key parameter of the assembly process is the determination of the assembly k-mer sizes, as it determines the assembler's capacity to handle the presence of organisms at different abundances in the sample and solve repetitive genomic regions. While large *k*-mer sizes handle repetitive regions in the genomes, shorter k-mers are best suited for the assembly of low-abundance taxa. For most datasets, using the default *k*-mer progression implemented in MEGAHIT (k = 21,31,41,51,61,71,81,91,99), which has been optimized through extensive testing on diverse metagenomic datasets, is recommended. The choice of the minimum contig length is a second important user-determined parameter. Short contigs (<1 kb) may be informative for specific analysis, whereas for MAG reconstruction, a minimum length of 2.5 kb or more is more appropriate, as longer contigs allow for a more reliable binning process. Importantly, specific computational methods should be considered for the assembly of long reads or the hybrid assembly of long and short reads. These methods are covered in Chapter 10.

According to their study design and research objectives, users need to determine if the metagenomic assembly will be performed on individual samples or in a co-assembly. Assembling several metagenomes together in a co-assembly allows recovery of low-abundance organisms in the samples and can simplify comparative analysis. However, it can lead to an increase in assembly chimera, for highly divergent samples. A recommendation therefore is to use a co-assembly approach when working with time series or if the samples are expected to share a large proportion of organisms.

After assembly, the quality of the assembly can be evaluated using the classic N50, L50, and total assembly length. Assessing the proportion of reads that were left out from the assembly can also provide valuable insights regarding the assembly quality. Tools like meta-QUAST [19] provide assembly statistics while accounting for the metagenomic nature of the data.

2.3.2 Binning

The binning process allows the grouping together of contigs from the same genome into a MAG. Several signals can be used to automatically bin contigs, including sequence composition (e.g., tetranucleotide frequencies) or differential coverage patterns between samples. A vast number of automatic binning tools such as MetaBAT2 [20], CONCOCT [21], MaxBin2 [22], or SemiBin2 [23] are available each with their specific strengths and limitations (*see* **Note 4**).

State-of-the-art MAGs recovery pipelines such as the nf-core/mag pipeline [24], which implements the field's best practices, and the MetaWrap pipeline [25] use the ensemble approach to obtain more complete and less contaminated MAGs. It is possible to compare and combine the results of each individual binning tool using utilities like DAS Tool [26], allowing to leverage each binning tool's strength while limiting their individual weaknesses.

The success of the binning step depends on several key variables, including the assembly quality and contig length, but also the presence of multiple samples in the dataset. Indeed, to leverage differential coverage-based binning algorithms, it is critical to have multiple samples containing the genome of interest. Typically, larger datasets with varying differential coverage patterns will allow the generation of higher-quality MAGs.

2.3.3 Quality Assessment

A rigorous assessment of the MAGs' quality after binning is essential to exclude low-quality bins before downstream analysis. A standardized metric and threshold of MAG quality is the Minimum Information about Metagenome-Assembled Genome (MIMAG) standards [27] which defines categories of MAG quality based on bin completeness and contamination. CheckM2 [28] is a popular choice to perform this MAG quality assessment for prokaryotic genomes, while BUSCO [29] or EukCC [30] allows the assessment of eukaryotic genomes. Importantly, the tool assesses the genome completeness and contamination using the presence and number of lineage-specific single-copy marker genes.

Following the MIMAG quality standard, bins of >90% completeness and < 5% contamination are considered high quality, while bins between 50 and 90% completeness and < 5% contamination are classified as medium quality draft. These metrics, while useful, may not completely capture the MAG quality when dealing with closely related species. In these situations, GUNC (Genome UNCluster) [31] can be used in conjunction with CheckM2, to detect any possible chimerism in MAGs. GUNC will assess the taxonomic consistency of all the genes in the MAG and may help detect the mixture of contigs from closely related species in a single bin. Finally, the detection of rRNA genes and tRNAs may also help to assess the bin quality, although these genes are often fragmented or missing due to their repetitive nature, which makes them challenging to capture using short-read metagenomics approaches. Tools like Barrnap (https://github.com/tseemann/barrnap) can be used to detect rRNA genes, while tRNAscan-SE [32] can identify and evaluate tRNA completeness. The presence and quality of these genes are required to meet the high-quality criteria of the MIMAG standard and can be useful for certain downstream analysis.

While not strictly part of the quality assessment process, the taxonomic classification of the MAGs can help detect contaminated or low-quality MAGs. Tools like GTDB-Tk can indeed suggest a potential taxonomic classification for the bin while also flagging potentially problematic MAGs that do not fit well within known taxonomic space.

2.3.4 Dereplication and Refinement

To generate a non-redundant collection of high-quality MAGs, several refinement steps are necessary. First, a dereplication step of highly similar MAGs to avoid inflating the true genomic diversity captured in each set of metagenomes. This step includes the within-sample dereplication when using different binning algorithms and the cross-sample dereplication to cluster similar genomes recovered from different samples. Tools such as dRep [33] compute pair-wise average nucleotide identity (ANI) between MAGs in a scalable manner. The choice of similarity thresholds depends on the study objectives with 95% ANI commonly used to approximate species-level grouping. However, studies that require a strain-level resolution could use higher thresholds.

Implementing an additional manual refinement step is recommended when trying to retrieve bins of interest. The Anvi'o platform [34] provides an interactive visualization framework for manual MAG curation and allows for easy integration of multiple genomic signals, including coverage patterns, GC content, taxonomic classifications, and completion estimates. This additional manual refinement is critical to identify and resolve common binning errors such as incorrectly assigned contigs, contamination from closely related species, and strain-level variations that automated binners might miss.

2.4 Generating MAGs and Gene Catalogs with MG-TK

MG-TK (metagenomic toolkit) is an open-source pipeline developed at the Quadram Institute, available from (https://github.com/hildebra/mg-tk). It was programmed to process raw metagenomic sequences to derive taxonomic and functional abundances, either via an assembly-dependent or assembly-free approach. In the assembly-dependent approach, MAGs are reconstructed and clustered to MGS (metagenomic species), and for each MGS, an intraspecific phylogeny is created to achieve strain-resolved metagenomics.

2.4.1 Assembly-Free Processing of Metagenomes

MG-TK is a pipeline for processing of short- and long-read WMS datasets. It can be run in a reference-based mode, which does not require an assembly and is most useful for environmental samples with high microbial diversity and lower sequencing coverage. In this case, after read filtering, the community composition is determined from the sequence data using a miTag approach [35]. The reads corresponding to rRNA subunits are extracted with SortMeRNA [36] and then searched against the SILVA database [37]. Reads matching the SSU and LSU databases are further filtered with custom scripts, using FLASH [38] to attempt merging all matched read pairs. If read pairs cannot be merged, single reads are interleaved such that the second read pair is reverse-complemented and sequentially added to the first read. Next, candidate interleaved or merged reads are fine-matched to the SILVA database using Lambda3 [39]. Based on these, the identity of reads is determined with the lowest common ancestor (LCA) algorithm adapted from LotuS [40] (*see* **Note 5**).

To determine functional profiles, the raw metagenomic reads are quality filtered and aligned via diamond [41] against common functional databases such as eggnog [42], CAZy [43], or KEGG [44]. Hits against these databases are quality filtered based on % identity, %query, and reference coverage, to determine the number of reads aligning to either database. The complete procedure for the functional and miTag approach is described in more detail in [45].

2.4.2 Generating Assemblies and Genomes from Metagenomes

MG-TK also provides a mode for both long- and short-read WMS, which in addition to implementing many of the steps discussed in this chapter includes strain level resolution of dereplicated MAGs.

In assembly mode, MG-TK carries out quality control, assembly, and binning, followed by dereplication into metagenomic species (MGS) shown in Fig. 2. A key difference compared to other pipelines is in the dereplication step, where the process for generating MGS combines both MAGs and canopy clusters of genes and is based on sharing of the GTDB-established 120 (bacterial) or

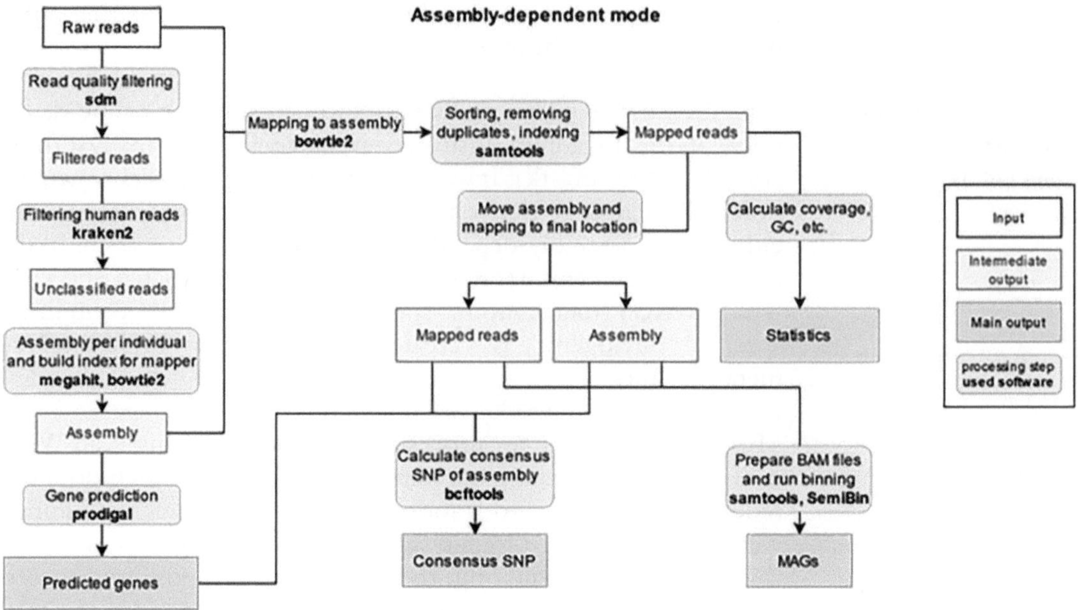

Fig. 2 Typical workflow in MG-TK running using an assembly-dependent mode. The default process for short reads is depicted. First, human reads are removed using either Kraken2 or Hostile, and the remaining reads are assembled (or co-assembled) using MEGAHIT. Filtered reads are mapped back to the assembly using Bowtie2 to generate coverage for each assembled contig. Genes are predicted from the assemblies using Prodigal, and subsequently clustered at 95% identity using Mmseqs2 to produce a gene catalog. Clusters of genes from this catalog are likely to originate from the same genome identified using canopy clustering, and MAGs are produced for each assembly using SemiBin2. Canopy clusters and MAGs are then grouped into MGS with a custom clustering algorithm using marker gene information and MAG quality. Abundance of an MGS is provided based on the coverage of its conserved marker genes in the gene catalog

40 (archaeal) marker genes, rather than ANI-like comparison. Including canopy clusters allows genes for which high-quality MAGs could be recovered to be associated with the MGS, expanding further the recovered diversity.

The primary results of the assembly-based mode are MAGs and MGS, as well as the taxonomic abundance of samples. Additionally, which genes from the gene catalog are present in each MAG are outputted, facilitating subsequent pan-genomic analysis.

2.4.3 Representing Gene Diversity in Samples Through a Gene Catalog

The gene catalog is a core part of many MG-TK processes and is created by clustering all the predicted genes from assemblies above 95% identity using mmSeqs2 [46]. The abundance of each gene is calculated for each sample based on coverage from reads aligned to the assemblies. Genes which are homologous to GTDB marker genes are located allowing taxonomic profiling and genome-resolved metagenomics as described above. Function is assigned to the genes by searching against multiple references, from broad (e.g., KEGG or eggnog) to specialized references (e.g., CAZy) or gut-specific BSB modules. These functional predictions along with the gene abundance allow the functional profiling of the whole community, as well as linking the gene content of individual MAGs to function.

2.4.4 Strain-Resolved Metagenomics

MGS can be used to obtain strain-level resolution. MG-TK uses consensus SNPs calculated for each sample with bcftools [47] to reconstruct an exact DNA sequence for each sample. Within each MGS, core genes are de novo determined by comparing different MAGs, and from highly-prevalent core genes, an intraspecific phylogeny is calculated, using mafft [48] or MUSCLE5 [49] for multiple sequencing alignments and iqTree2 [50] or RAxML [51] to reconstruct a maximum likelihood phylogeny. This approach allows the reduction of the reference bias and avoids inflating the mapping space. Importantly, MG-TK assumes the presence of one dominant strain per species per metagenome. Therefore, sequences arising from conspecific strains, i.e., multiple strains of the same species within a given metagenome, are filtered out based on the frequency of non-reference alleles in each contig. The phylogenetic trees can then be used, for example, to analyze strain sharing between individuals or to identify strains associated with certain conditions, as we recently used to track the sharing of gut microbes among family members [52].

3 Notes

1. There is an important trade-off between complete host read exclusion and removal of potential microbial sequences. Aggressive filtering may provide a more thorough host DNA removal, but risk eliminating true microbial sequences that share similarity with host genomic regions. Conversely, lenient filtering may preserve more microbial information but at the cost of increased host contamination.
2. Random subsampled data during initial exploratory analyses can be used if computational limitations are a concern.
3. In case of limited memory availability, MEGAHIT is better suited than metaSPADES which has higher memory requirements but can generate longer contigs.
4. No single binning algorithm consistently outperforms the others, and combining several binning tools together has emerged as a best practice for high-quality MAG recovery.
5. This approach is suitable for profiling Bacteria, Fungi, Archaea, and Eukaryotes.

Acknowledgements

This work was funded by the Biotechnology and Biological Sciences Research Council (BBSRC) through an Institute Strategic Programme award to the QIB Programmes Gut Microbes and Health BB/R012490/1 and associated Core Capability Grant BB/CCG1860/1 and Food, Microbiome and Health [BB/X011054/1] and associated Core Capability Grant BB/CCG2260/1.

References

1. Gihawi A, Ge Y, Lu J et al (2023) Major data analysis errors invalidate cancer microbiome findings. MBio 14:e01607-23. https://doi.org/10.1128/mbio.01607-23
2. Constantinides B, Hunt M, Crook DW (2023) Hostile: accurate decontamination of microbial host sequences. Bioinformatics 39:btad728. https://doi.org/10.1093/bioinformatics/btad728
3. Chen S (2023) Ultrafast one-pass FASTQ data preprocessing, quality control, and deduplication using fastp. iMeta 2:e107. https://doi.org/10.1002/imt2.107
4. Wood DE, Lu J, Langmead B (2019) Improved metagenomic analysis with kraken 2. Genome Biol 20:257. https://doi.org/10.1186/s13059-019-1891-0
5. Kim D, Song L, Breitwieser FP, Salzberg SL (2016) Centrifuge: rapid and sensitive classification of metagenomic sequences. Genome Res 26:1721–1729. https://doi.org/10.1101/gr.210641.116
6. Ounit R, Wanamaker S, Close TJ, Lonardi S (2015) CLARK: fast and accurate classification of metagenomic and genomic sequences using discriminative k-mers. BMC Genomics 16:236. https://doi.org/10.1186/s12864-015-1419-2
7. Blanco-Míguez A, Beghini F, Cumbo F et al (2023) Extending and improving metagenomic taxonomic profiling with uncharacterized species using MetaPhlAn 4. Nat Biotechnol 41:1633–1644. https://doi.org/10.1038/s41587-023-01688-w
8. Milanese A, Mende DR, Paoli L et al (2019) Microbial abundance, activity and population genomic profiling with mOTUs2. Nat Commun 10:1014. https://doi.org/10.1038/s41467-019-08844-4
9. Thomas AM, Segata N (2019) Multiple levels of the unknown in microbiome research. BMC Biol 17:48. https://doi.org/10.1186/s12915-019-0667-z
10. Lu J, Breitwieser FP, Thielen P, Salzberg SL (2017) Bracken: estimating species abundance in metagenomics data. PeerJ Comput Sci 3:e104. https://doi.org/10.7717/peerj-cs.104
11. Wright RJ, Comeau AM, Langille MGI (2023) From defaults to databases: parameter and database choice dramatically impact the performance of metagenomic taxonomic classification tools. Microb Genom 9:000949. https://doi.org/10.1099/mgen.0.000949
12. Chaumeil P-A, Mussig AJ, Hugenholtz P, Parks DH (2022) GTDB-Tk v2: memory friendly classification with the genome taxonomy database. Bioinformatics 38:5315–5316. https://doi.org/10.1093/bioinformatics/btac672
13. Meyer F, Fritz A, Deng Z-L et al (2022) Critical assessment of metagenome interpretation: the second round of challenges. Nat Methods 19:429–440. https://doi.org/10.1038/s41592-022-01431-4
14. Meyer F, Bremges A, Belmann P et al (2019) Assessing taxonomic metagenome profilers with OPAL. Genome Biol 20:51. https://doi.org/10.1186/s13059-019-1646-y
15. Beghini F, McIver LJ, Blanco-Míguez A et al (2021) Integrating taxonomic, functional, and strain-level profiling of diverse microbial communities with bioBakery 3. elife 10:e65088. https://doi.org/10.7554/eLife.65088
16. Zheng J, Ge Q, Yan Y et al (2023) dbCAN3: automated carbohydrate-active enzyme and substrate annotation. Nucleic Acids Res 51:W115–W121. https://doi.org/10.1093/nar/gkad328
17. Li D, Liu C-M, Luo R et al (2015) MEGAHIT: an ultra-fast single-node solution for large and complex metagenomics assembly via succinct de Bruijn graph. Bioinformatics 31:1674–1676. https://doi.org/10.1093/bioinformatics/btv033
18. Nurk S, Meleshko D, Korobeynikov A, Pevzner PA (2017) metaSPAdes: a new versatile

metagenomic assembler. Genome Res 27:824–834. https://doi.org/10.1101/gr.213959.116

19. Mikheenko A, Saveliev V, Gurevich A (2016) MetaQUAST: evaluation of metagenome assemblies. Bioinformatics 32:1088–1090. https://doi.org/10.1093/bioinformatics/btv697
20. Kang DD, Li F, Kirton E et al (2019) MetaBAT 2: an adaptive binning algorithm for robust and efficient genome reconstruction from metagenome assemblies. PeerJ 7:e7359. https://doi.org/10.7717/peerj.7359
21. Alneberg J, Bjarnason BS, de Bruijn I et al (2014) Binning metagenomic contigs by coverage and composition. Nat Methods 11: 1144–1146. https://doi.org/10.1038/nmeth.3103
22. Wu Y-W, Simmons BA, Singer SW (2016) MaxBin 2.0: an automated binning algorithm to recover genomes from multiple metagenomic datasets. Bioinformatics 32:605–607. https://doi.org/10.1093/bioinformatics/btv638
23. Pan S, Zhao X-M, Coelho LP (2023) SemiBin2: self-supervised contrastive learning leads to better MAGs for short- and long-read sequencing. Bioinformatics 39:i21–i29. https://doi.org/10.1093/bioinformatics/btad209
24. Krakau S, Straub D, Gourlé H et al (2022) nf-core/mag: a best-practice pipeline for metagenome hybrid assembly and binning. NAR Genomics Bioinformatics 4:lqac007. https://doi.org/10.1093/nargab/lqac007
25. Uritskiy GV, DiRuggiero J, Taylor J (2018) MetaWRAP – a flexible pipeline for genome-resolved metagenomic data analysis. Microbiome 6:158. https://doi.org/10.1186/s40168-018-0541-1
26. Sieber CMK, Probst AJ, Sharrar A et al (2018) Recovery of genomes from metagenomes via a dereplication, aggregation and scoring strategy. Nat Microbiol 3:836–843. https://doi.org/10.1038/s41564-018-0171-1
27. Bowers RM, Kyrpides NC, Stepanauskas R et al (2017) Minimum information about a single amplified genome (MISAG) and a metagenome-assembled genome (MIMAG) of bacteria and archaea. Nat Biotechnol 35: 725–731. https://doi.org/10.1038/nbt.3893
28. Chklovski A, Parks DH, Woodcroft BJ, Tyson GW (2023) CheckM2: a rapid, scalable and accurate tool for assessing microbial genome quality using machine learning. Nat Methods 20:1203–1212. https://doi.org/10.1038/s41592-023-01940-w
29. Seppey M, Manni M, Zdobnov EM (2019) BUSCO: assessing genome assembly and annotation completeness. In: Kollmar M (ed) Gene prediction: methods and protocols. Springer, New York, pp 227–245
30. Saary P, Mitchell AL, Finn RD (2020) Estimating the quality of eukaryotic genomes recovered from metagenomic analysis with EukCC. Genome Biol 21:244. https://doi.org/10.1186/s13059-020-02155-4
31. Orakov A, Fullam A, Coelho LP et al (2021) GUNC: detection of chimerism and contamination in prokaryotic genomes. Genome Biol 22:178. https://doi.org/10.1186/s13059-021-02393-0
32. Chan PP, Lowe TM (2019) tRNAscan-SE: searching for tRNA genes in genomic sequences. In: Kollmar M (ed) Gene prediction: methods and protocols. Springer, New York, pp 1–14
33. Olm MR, Brown CT, Brooks B, Banfield JF (2017) dRep: a tool for fast and accurate genomic comparisons that enables improved genome recovery from metagenomes through de-replication. ISME J 11:2864–2868. https://doi.org/10.1038/ismej.2017.126
34. Eren AM, Esen ÖC, Quince C et al (2015) Anvi'o: an advanced analysis and visualization platform for omics data. PeerJ 3:e1319. https://doi.org/10.7717/peerj.1319
35. Salazar G, Ruscheweyh H-J, Hildebrand F et al (2021) mTAGs: taxonomic profiling using degenerate consensus reference sequences of ribosomal RNA genes. Bioinformatics 38: 270–272. https://doi.org/10.1093/bioinformatics/btab465
36. Kopylova E, Noé L, Touzet H (2012) SortMeRNA: fast and accurate filtering of ribosomal RNAs in metatranscriptomic data. Bioinformatics 28:3211–3217. https://doi.org/10.1093/bioinformatics/bts611
37. Quast C, Pruesse E, Yilmaz P et al (2013) The SILVA ribosomal RNA gene database project: improved data processing and web-based tools. Nucleic Acids Res 41:D590–D596. https://doi.org/10.1093/nar/gks1219
38. Magoč T, Salzberg SL (2011) FLASH: fast length adjustment of short reads to improve genome assemblies. Bioinformatics 27:2957–2963. https://doi.org/10.1093/bioinformatics/btr507
39. Hauswedell H, Hetzel S, Gottlieb SG et al (2024) Lambda3: homology search for protein, nucleotide, and bisulfite-converted sequences. Bioinformatics 40:btae097.

https://doi.org/10.1093/bioinformatics/btae097

40. Özkurt E, Fritscher J, Soranzo N et al (2022) LotuS2: an ultrafast and highly accurate tool for amplicon sequencing analysis. Microbiome 10:176. https://doi.org/10.1186/s40168-022-01365-1
41. Buchfink B, Reuter K, Drost H-G (2021) Sensitive protein alignments at tree-of-life scale using DIAMOND. Nat Methods 18:366–368. https://doi.org/10.1038/s41592-021-01101-x
42. Huerta-Cepas J, Szklarczyk D, Heller D et al (2019) eggNOG 5.0: a hierarchical, functionally and phylogenetically annotated orthology resource based on 5090 organisms and 2502 viruses. Nucleic Acids Res 47:D309–D314. https://doi.org/10.1093/nar/gky1085
43. Cantarel BL, Coutinho PM, Rancurel C et al (2009) The carbohydrate-active EnZymes database (CAZy): an expert resource for Glycogenomics. Nucleic Acids Res 37:D233–D238. https://doi.org/10.1093/nar/gkn663
44. Kanehisa M, Goto S (2000) KEGG: Kyoto encyclopedia of genes and genomes. Nucleic Acids Res 28:27–30. https://doi.org/10.1093/nar/28.1.27
45. Bahram M, Hildebrand F, Forslund SK et al (2018) Structure and function of the global topsoil microbiome. Nature 560:233–237. https://doi.org/10.1038/s41586-018-0386-6
46. Steinegger M, Söding J (2017) MMseqs2 enables sensitive protein sequence searching for the analysis of massive data sets. Nat Biotechnol 35:1026–1028. https://doi.org/10.1038/nbt.3988
47. Danecek P, Bonfield JK, Liddle J et al (2021) Twelve years of SAMtools and BCFtools. GigaScience 10:giab008. https://doi.org/10.1093/gigascience/giab008
48. Katoh K, Misawa K, Kuma K, Miyata T (2002) MAFFT: a novel method for rapid multiple sequence alignment based on fast Fourier transform. Nucleic Acids Res 30:3059–3066
49. Edgar RC (2022) Muscle5: high-accuracy alignment ensembles enable unbiased assessments of sequence homology and phylogeny. Nat Commun 13:6968. https://doi.org/10.1038/s41467-022-34630-w
50. Minh BQ, Schmidt HA, Chernomor O et al (2020) IQ-TREE 2: new models and efficient methods for phylogenetic inference in the genomic era. Mol Biol Evol 37:1530–1534. https://doi.org/10.1093/molbev/msaa015
51. Stamatakis A (2014) RAxML version 8: a tool for phylogenetic analysis and post-analysis of large phylogenies. Bioinformatics 30:1312–1313. https://doi.org/10.1093/bioinformatics/btu033
52. Hildebrand F, Gossmann TI, Frioux C et al (2021) Dispersal strategies shape persistence and evolution of human gut bacteria. Cell Host Microbe 29:1167–1176.e9. https://doi.org/10.1016/j.chom.2021.05.008

Chapter 12

Statistical Analysis of Microbiome Data

George M. Savva and Alise J. Ponsero

Abstract

The statistical analysis of microbiome data presents unique challenges including high dimensionality, compositional constraints, sparsity, and phylogenetic structure that complicate conventional statistical approaches. This chapter outlines key steps for common microbiome analysis tasks including research question formulation, data preprocessing, descriptive analysis, statistical modeling, and validation for comparisons of diversity and differential abundance estimation. Specific methods are discussed, as well as general considerations for selecting methods and avoiding common statistical pitfalls in this rapidly evolving field.

Key words Microbial diversity, Differential abundance, Diversity, Normalization, Statistical analysis

1 Introduction

The appropriate analysis of microbiome data depends on the research question, study design, and the characteristics of the dataset. This chapter outlines key principles and practical tips to support each step of the analysis process. Analysis of microbiome data should align with the usual principles of good statistical analysis, but there are specific features of microbiome data that can complicate its analyses and interpretation. These are as follows:

High dimension: Datasets arising from microbiomes have many features (corresponding to microbial taxa, genes, or functions) but relatively few samples.

Count data: Microbiome data are usually generated as counts of reads that have been assigned to each feature, and some preprocessing is usually required before they can be statistically modeled.

Compositionality: The total number of reads obtained per sample is often fixed, or is a technical artefact not directly related to microbial load. So absolute counts have little meaning without

Simon R. Carding (ed.), *Best Practice in Microbiome Research*, Springer Protocols Handbooks,
https://doi.org/10.1007/978-1-0716-5009-7_12, © The Author(s) 2026

external calibration, and microbial datasets should be regarded as compositional.

Sparsity and overdispersion: Microbial data are likely to be sparse, with many low and zero counts, alongside fewer extremely large counts.

Phylogenetic structure: Features are related, and reads may be classified at different phylogenetic ranks or in gene/pathway clusters.

Missing data: Many reads will not be classifiable at the desired taxonomic rank, particularly when attempting to map reads to individual species.

There is no single correct method for handling these issues; numerous analytical approaches exist, each with its own strengths, assumptions, limitations, and suitability in different contexts. As such, analysts must make informed choices about data processing and statistical modeling. These decisions should be guided by the characteristics of the data and aims of the project, general principles of sound statistical practice, and empirical research findings.

Important principles to keep in mind are that an analysis should be:

1. *Informative:* directly addresses a research question.
2. *Valid*: respecting the structure of the data and the design of the study.
3. *Efficient*: extracting the maximum reliable information from available data.
4. *Understandable and acceptable*: communicating clearly to the intended audience, using methods accepted in the field.
5. *Robust*: with consistent results under reasonable variations in modeling approach.
6. *Pre-specified*: planned through a detailed, written analysis protocol.
7. *Reproducible*: implemented using transparent, well-documented code or pipelines.

2 Methods

Statistical analysis of microbiome data involves the following phases:

1. *Setting the question and planning*: clearly articulating precise questions that you will answer, and which analyses will address them.

2. *Preprocessing*: creating analysis datasets from raw data, through data cleaning, normalization, and combining data and metadata from different sources.
3. *Descriptive analysis and visualization*: describing the samples and creating graphics that help to highlight important features of the dataset.
4. *Modeling*: using statistical models to quantitatively answer scientific questions.
5. *Validation*: ensuring that the models you have used are appropriate for the dataset.

2.1 Defining the Research Question and Establishing a Statistical Analysis Plan

Before beginning an analysis:

- Define clear research questions and hypotheses.
- Identify appropriate statistical outputs for addressing these questions.
- Select suitable methods and software packages based on study design and available data and resources.
- Review limitations and assumptions of chosen statistical methods.
- Document all decisions around analysis methods in a statistical analysis plan.

2.1.1 Defining the Research Question

In gut microbiome research, research questions often concern associations between aspects of the microbiome (i.e., its composition or function) and other features, such as demographics, environmental exposures, treatments, treatment effects, biomarkers, or disease outcomes. A specific microbial feature of interest might be targeted, such as the abundance of a specific microbial taxon, function, or metabolic reaction, or they might be exploratory, such as aiming to discover which microbial features might cause or be caused by other factors.

More complex questions might concern microbe/microbe interactions, or the microbiome as a mediator, moderator, a predictor, or a diagnostic tool. In all cases, the research question(s), both targeted and exploratory, should be specified in as much detail as possible.

Questions focused solely on associations are vague and should be avoided; instead, consider whether the underlying question is causal or predictive in nature. The answers to causal questions will inform possible interventions, whereas predictive questions will inform prognosis or diagnosis, and different statistical analyses will be needed for each.

While it may not be possible to directly identify causal relationships in many research designs, particularly observational designs, thinking causally is still helpful to design analyses, in particular to identify important covariates and statistical methods to reduce bias from potential confounding factors, or to avoid bias from conditioning on colliders [1]. Consider using a directed acyclic graph to identify the causal relationships between the microbes and aspects of the metadata [2].

If the research concerns prediction, then machine learning methods might be preferred, and the results will be judged differently, usually in terms of the sensitivity and specificity of a diagnostic test, or prediction accuracy. In this case, whether an identified relationship is causal or not is of less interest, but internal cross-validation or validation with external datasets should be used to confirm predictive power.

2.1.2 Statistical Analysis Plan

A statistical analysis plan should include [3] the following:

1. The specific research questions, their rationale, and definitions of outcome measures.
2. An outline of the study design.
3. Approaches used to assess data integrity.
4. Methods for preprocessing.
5. Analyses that will address individual objectives, including specific statistical models, transformations, and covariates.
6. Methods to account for missing data and multiplicity if needed.
7. Validation strategies and sensitivity analyses.

Statistical plans for microbiome analyses are particularly helpful given the number of possible statistical approaches that it is possible to take. They should be agreed upon within a research team, to agree on the research questions and the outcomes, and so that the correct resources can be allocated. Statistical analysis plans can be adapted once analysis has begun, but these changes should be documented.

2.2 Preprocessing the Dataset

The result of a bioinformatic analysis of sequencing data is a table of microbial feature abundances for each sample, and a corresponding table of "metadata" that describes the other characteristics of each sample.

Microbial features are likely to be individual microbial taxa, or genes, and are usually expressed as counts, corresponding to the number of times each feature was detected in the sample. Some preprocessing is usually necessary before statistical analysis, but this depends on the planned analysis, and it is important to understand the requirements for each downstream method.

Normalization ensures that datasets from different samples can be fairly compared. *Transformations* are used to improve the validity of statistical methods, *decontamination* removes reads that are not from the parent population of interest, and *filtering* and

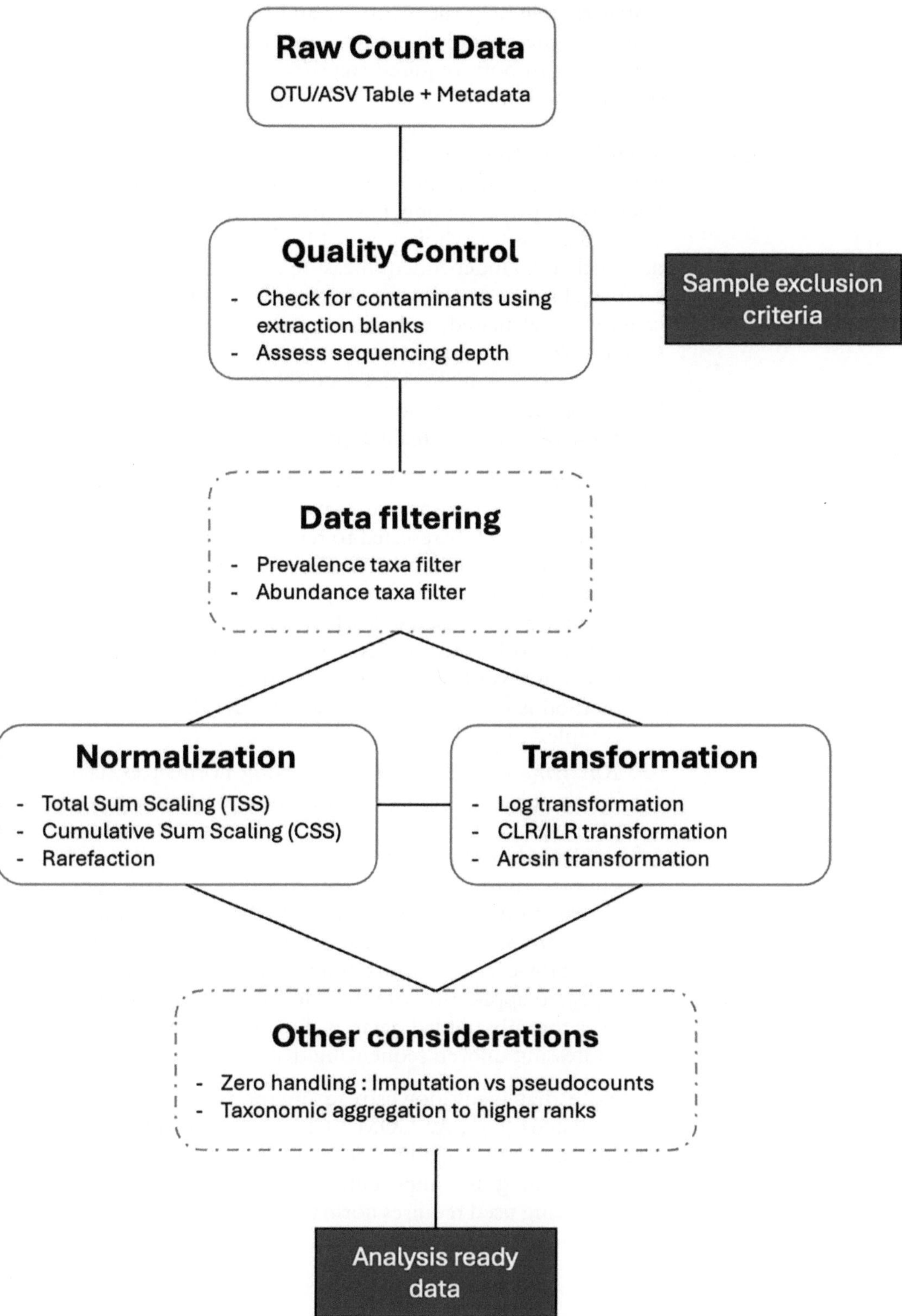

Fig. 1 Overview of a standard preprocessing workflow for microbiome data analysis

agglomeration help the feasibility and power of analyses. Figure 1 illustrates the progression from raw count data to analysis-ready datasets, with both required steps (solid blue boxes) and optional steps (dashed boxes).

2.2.1 Data Normalization

Microbiome data is typically compositional, with the total count within a sample determined by sampling and sequencing processes. Each count therefore only has a meaning as a fraction of the reads assigned across the whole sample [4, 5]. Without another source of data such as an independent measure of the total bacterial load in a sample, absolute abundances of microbes in the parent population cannot be calculated, only their *relative abundances* within each sample. Normalization focuses on enabling comparisons between samples with potentially different total counts. Many methods have been proposed for the normalization of microbiome data [6, 7], and these fall into two main approaches:

1. *Scaling methods*: The most common approaches are total sum scaling (TSS) or cumulative sum scaling (CSS). With TSS, counts are simply rescaled to represent their proportion within a sample (their relative abundances). CSS is a generalization of this approach [8], in that a proportion of the most abundant features are not included when calculating the total. The proportion to be included is determined based on matching the distributions of the remaining features between samples. This method is more robust to large outliers than TSS but is more complex to implement.
2. *Rarefying and rarefaction:* Rarefying normalizes data by randomly subsampling an equal number of counts from all samples, typically based on the size of the smallest total count. It is particularly useful prior to comparisons of diversity between samples. Rarefying has been criticized as large amounts of information can be lost, particularly for rare features, or if there are samples in a dataset with very small total counts [9]. However, a correct implementation of rarefaction, the repeated application of rarefying followed by pooling statistics over the subsampled datasets, has been shown to be robust to addressing uneven sequencing depth [7].

Note that many downstream methods for differential abundance analysis (e.g., ANCOM-BC2, MaAslin3) do use count data directly, implicitly including normalization during or prior to statistical modeling. It is important to understand whether the analysis method being used requires normalized or unnormalized data as an input.

2.2.2 Data Transformations

Following normalization, data may still need to be transformed before statistical analysis. When selecting a transformation, consider the distribution of data, the requirements of the downstream analysis methods, and how results using the transformed datasets should be interpreted.

Simple mathematical transformations (logarithmic or arcsin square root) are used for reducing the extreme skew that is typical of relative abundances and for stabilizing variances across groups. These enable analyses based on the assumptions of normality and homoskedasticity to be conducted.

Zeros are common in microbiome data, so applying logarithms to abundances directly is often not possible. A common approach that is applied automatically by several packages is to add a small constant (pseudo-count) before log-transformation. Where there is a high proportion of zeros, the choice of pseudo count can influence the analysis, and so sensitivity analyses should be applied to make sure that results are not dependent on this arbitrary choice.

Pseudo-counts should be applied after normalization so that their impact is constant across samples. Logarithms are also sensitive to very small values which might be imprecisely measured; their magnitude hence the magnitude of their measurement error is vastly inflated under a log-transform. Hence, log transforms should be avoided with very sparse data.

The arcsin-square root transformation is sometimes used instead. This maps relative abundances from the interval $[0,1]$ onto a bounded range $[0,\pi/2]$, while naturally handling proportions at the extremes. This transformation is particularly useful when working with relative abundances, as it tends to stabilize the variance of proportional data and handles zeros without modification.

2.2.3 Log-Ratios for Compositional Data

Differences in microbial relative abundances can be difficult to interpret because changes in the relative abundance of a feature can reflect changes anywhere in the microbiota.

Transformation with log-ratios (the logarithm of the ratios between taxa or groups of taxa) can reduce this dependence between microbial feature counts, meaning that changes in individual scores can be more directly interpreted.

While relative abundances of microbial features are affected by the compositional nature of data, the ratios (and hence log-ratios) between different features are not. That is, given three taxa A, B, and C, while the relative abundances of A and B will both be affected by changes in the abundance of taxon C, the ratio of the abundances of A and B will not. Hence, the ratio of taxon abundances can more easily be compared between samples than the abundances themselves. There are three main algorithms for applying this idea to a dataset:

The *additive log ratio (ALR)* transformation proceeds by choosing a reference feature and then computing the log-ratio of each feature relative to this reference.

Centered log ratios are the most commonly applied, and avoid the need to select a particular reference by using the geometric mean of all features instead. Hence, a CLR normalizes each taxon abundance by the geometric average of all taxon abundances, thereby providing some robustness against large shifts in a single taxon affecting the apparent abundance of the others. CLRs can be affected by zeros, since the geometric mean is undefined if any individual terms are zero, and they are sensitive to the choice of pseudo-count. Hence, CLRs are best applied to more filtered or agglomerated datasets [10] and should be applied following normalization such as TSS (this pre-normalization is often done automatically by software packages).

Finally, the *isometric log-ratio (ILR)* relies on the selection of binary partitions of the features in the dataset that are of analytical interest, and then finding the log-ratio between abundances representing each of these partitions. For microbiome data, particularly, the PhILR (phylogenetic ILR) was proposed to incorporate phylogenetic relationships between taxa into the log-ratio transformation, with each balance representing the partition defined by the edge of a phylogenetic tree [11]. This approach both handles the compositionality of microbiome data and reduces the dimension in a biologically meaningful way by grouping phylogenetically similar organisms.

2.2.4 Count-Based Methods for Functional Data

When analyzing functional data (genes or pathways) or virome datasets (see Chap. 14 for viromics analysis), different data transformations need to be applied to allow the normalization by both the gene/sequence length and the sequencing depth. The following transformation can be considered:

1. CPM (counts per million) normalizes for sequencing depth.
2. RPKM (reads per kilobase million) additionally normalizes for gene length.
3. TPM (transcripts per million) provides a more consistent normalization across samples.

These normalizations are particularly important when comparing genes/pathways of different lengths or across samples with different sequencing depths.

2.2.5 Sparsity and Zero-Inflation

Rare features will often have small counts or zero counts across many samples. This presents several challenges. First, transformations often rely on logarithms or log-ratios, which cannot be directly used with zeros. Excess zeros can violate modeling assumptions of methods based on normal distributions or negative binomial count models.

Several statistical approaches to zeros have been applied to microbiome data. Zero-inflated models explicitly account for excess zeros in the data by modeling them as a mixture of true zeros (absence) and sampling zeros (failure to detect). Many differential abundance methods including ALDEx2 and ANCOM-BC2 handle zeros through their underlying statistical framework, making them particularly suitable for sparse microbiome data.

2.2.6 Independent Filtering

Conducting differential abundance tests of many very rare taxa can force a severe correction for multiplicity, but with no prospect of making genuine new discoveries, making it harder to detect true positive effects. So, it can be helpful to filter out features with a low prevalence or abundance, either prior to calculation or before applying a multiplicity correction:

1. Before analysis, remove all taxa where it is unlikely to be any chance of discovering differential abundance.
2. Be sure to select and apply this filter independently of any data other than the feature table itself; otherwise, there is a risk of introducing bias. Typical filters will exclude taxa that have a low coefficient of variation across samples, have a low average abundance, or are present in only a few samples.
3. Typically, taxa present in fewer than 5% or 10% of samples might be removed, but this should be considered carefully, based on the total sample size, and the potential importance of ignoring rare taxa vs the power to detect effects in more common taxa.

2.2.7 Handling the Unknown Fraction of the Metagenomes

Often a high proportion of sequences cannot be assigned to a specific taxon or functional cluster. These unclassified reads can represent a large fraction of the metagenomes, particularly when working in understudied ecosystems but even in human gut samples. The treatment of these unknowns can impact the downstream analyses. The most conservative approach is their exclusion before data normalization, but this approach may discard potentially relevant biological information.

A more nuanced approach involves assigning these reads to a higher taxonomic level where the classification may be more reliable.

How to handle unassigned reads should be guided by the specific context of your study. In samples where a large proportion of reads are unassigned, complete removal could substantially distort the compositional nature of the data. In such cases, keeping unassigned reads as a separate category for relative abundance calculations might be more appropriate. Whatever approach is chosen, it is important to document the decision and consider assessing its impact on the results through sensitivity analysis.

2.2.8 Decontamination

Contamination in microbiome studies can arise from multiple sources and, if not properly addressed, can lead to spurious conclusions. This is particularly problematic in low-biomass samples where contaminant DNA can comprise a significant proportion of the sequenced material.

The detection and removal of contaminants requires the inclusion of appropriate negative controls in the study design. Statistical approaches, such as those implemented in the decontam and MicrobIEM R packages [12], can help identify potential contaminants based on their pattern of occurrence across samples and negative controls.

2.2.9 Taxonomic Aggregation (Agglomeration)

Taxonomic aggregation combines read counts from lower taxonomic ranks such as species-level counts, to higher levels, such as genus, family, or phylum. This reduces data sparsity by grouping rare taxa with their relatives, decreases dimensionality (the number of independent features), reduces the unclassified fraction, and can mitigate classification errors that are more common at the species level. The optimal aggregation level depends on analysis-specific objectives, and the dataset characteristics should be considered in this decision, such as the sequencing approach (amplicon vs. WGS), classification quality, and sample size. The proportion of classified reads at each taxonomic level can guide this decision.

Analyses can be repeated at different levels of taxonomic aggregation, but care should be taken as this increases the risk of false-positive associations and should be adjusted for when interpreting results.

2.3 Exploratory and Descriptive Analysis of the Dataset

Exploratory data analysis (EDA) involves calculating simple statistics and making visualizations to summarize data. This can guide later analysis and help readers to understand the dataset. It is important not to conflate the descriptive information with inferential analysis, where relationships suggested by these visualizations should be formally tested (acknowledging the exploratory nature of this approach).

2.3.1 Key Statistics and Their Visualization

The primary and secondary objectives of the study, as agreed in the study design, should guide the selection of statistics and visualizations. There is no consensus regarding the statistics that best reflect gut health or microbiome health, but the most common statistics and visualizations used to provide a basic characterization of the dataset are as follows:

1. Lists of commonly occurring features and their relative abundances.

2. Measures of diversity between and within samples.
3. Graphical representations of these in the form of trees, ordination plots, bar charts, or heatmaps.

Transformed relative abundances of common features or features relevant to your specific analysis are a useful place to start. These can be listed or are commonly represented by heatmaps or barplots. Heatmaps also allow both the taxa and the samples to be clustered, accommodate transformations, and allow for metadata to be described, which can provide some additional information as to the experimental design. Stacked bar plots can enable comparisons of broad structure across groups, but are not useful for discerning many different taxa, or low-abundance taxa, and so are likely best applied at higher taxonomic ranks.

2.3.2 Alpha Diversity

Diversity measures are summary statistics reflecting overall community composition and structure. Alpha diversity reflects microbial diversity within individual samples, while beta diversity measures diversity or dissimilarity between samples.

Different alpha diversity measures reflect both the "richness" of a sample (how many different taxa are present at a given taxonomic rank) and its "evenness" (whether a small number of taxa dominate). Higher diversity is generally thought of as a positive health outcome, although this is not always true [13].

Commonly used alpha diversity metrics are as follows:

1. *Microbial richness*: an estimate of the number of different taxa found within a sample.
2. *Simpson's index*: the probability that two individuals from the population, selected at random, belong to the same taxon. The Simpson's index is largely determined by the evenness of the more dominant taxa in a population and is less sensitive to the number or abundance of rare taxa.
3. *Shannon diversity*: captures both richness and evenness, giving more weight to rare species than Simpson's index. It increases with both the number of unique species and the evenness of their distribution.

Providing and comparing multiple measures across samples can be informative, since they reflect different aspects of microbial diversity.

2.3.3 Beta-Diversity and Associated Visualization

While "alpha diversity" measures the diversity within individual samples, "beta diversity" reflects the difference in the compositions of a pair of samples. Beta diversity is used to describe how related different sets of samples are, so it can be used to describe or measure the differences between and within groups of samples.

Hence, within a set of N samples, beta diversity is represented as an $N \times N$ matrix where the ij-th element is the dissimilarity between the i-th and the j-th sample.

As with alpha diversity, different measures of dissimilarity between samples exist. These can broadly be classified as follows:

1. *Incidence based:* Analogous to microbial richness, these rely on simply comparing the overlap in the presence or absence of taxa amongst samples. These measures are very sensitive to misclassification of small numbers of reads; they discard abundance information, are biased when sampling is incomplete, or sequencing depth varies considerably across samples. The Jaccard distance is a commonly used incidence-based distance.
2. *Abundance based:* These give more weight to the more abundant taxa in a dataset, and so better reflect differences in the overall compositions of microbiomes. They are less sensitive to errors in small numbers of reads and to rare features being undersampled. Bray-Curtis distance is the most widely used abundance-based measure [14].
3. *Compositional distances:* Euclidean distance can be calculated between sample pairs following log-ratio transformations, considering the data compositionality (known as Aitchison distance if formed after CLR). Such distances are more sensitive to the abundance of rare taxa than the Bray-Curtis distance, and so reflect a different aspect of the microbial composition.
4. *Phylogenetic distances:* Each of the above distances ignores the phylogenetic relatedness of different taxa. Other diversity measures can incorporate phylogenetic information into their comparison of samples. The most commonly applied example is the UniFrac distance, which exists in unweighted (ignoring abundance data) and weighted (using abundance data) forms [15].

To visualize beta-diversities between and within population subgroups, the dissimilarity matrix must be represented in a form that can be represented in a two-dimensional plane.

Common approaches include ordination methods such as principal coordinate analysis (PCoA), which can reveal broad patterns in beta diversity, and non-metric multidimensional scaling (NMDS), which is particularly useful for identifying clusters or gradients in community composition. While principal components analysis (PCA) is not directly applicable to microbiome data because of the compositionality constraint, it can be used following the Aitchison distance calculation. In any case, the extent to which the distance matrix can be successfully represented in two dimensions should be assessed, for example, by measuring the "stress" for an NMDS ordination plot or the percent variance explained for PCA.

Distance-based redundancy analysis (dbRDA) extends these approaches by directly modeling how other factors explain the variation in community composition and allow users to identify factors that most strongly correlate with microbial community differences [16], by quantifying how much of the observed taxonomic variation can be attributed to specific measured variables.

2.4 Statistical Modeling and Testing

2.4.1 Comparing Alpha Diversity and One-Dimensional Statistics Between Samples

If each microbiome sample can be represented by one or more independent statistics, such as an alpha diversity measure, (transformed and normalized) taxon abundances, genes, function abundances, or a ratio of interest, then conventional statistical methods and models (for example, ANOVA, linear models, etc.) can be applied to estimate the relationships between these statistics and features.

The considerations for these will be the same as for any other application of these statistical methods. They must reflect the study design, incorporate appropriate covariates, and account for dependencies in the data, for example, the appropriate handling of repeated measures.

For normally distributed independent measurements with homogeneous variance, methods based on linear models should be used. However, microbiome-derived measures often violate these assumptions. When distribution assumptions cannot be met, extensions of these models including generalized linear models or mixed effects might be used, or nonparametric alternatives if they can accommodate the study design.

2.4.2 Testing Beta Diversity

Testing the statistical significance of any clustering identified using beta-diversity provides a statistical test of whether the average compositions of microbiomes differ in a general sense between groups. This can be done by testing the scores from, for example, PCA or NMDS ordination using simple tests, or by using permutation tests based on the beta-diversities themselves. The most commonly applied permutation test is PERMANOVA [17], which calculates what proportion of the beta diversity in a sample is attributable to between-group differences and then, analogous to ANOVA, compares this value to the distribution that would be expected if there was no true group difference.

Like ANOVA, PERMANOVA relies on the assumption of equality of dispersions between groups and the exchangeability of the data points being permuted. In practical terms, exchangeability means that permutations must take place in blocks and strata that reflect the study design, and that repeated measures from experimental units are kept together as plots. When complex experimental designs are used, a naive application of PERMANOVA will lead to invalid results, and restricted permutations must be used [18]. Where the equality of dispersions is questionable, PERMANOVA may return false positives, and so alternative methods should be used, or results interpreted with caution.

2.4.3 Differential Abundance Analysis

Differential abundance analysis is an exploratory analysis process of building a statistical model for each feature of the microbiome, estimating the association between each taxon and other sample characteristics of interest, then identifying those for which the association is statistically significant, while controlling the false-positive rate.

As described above, it is possible to develop a linear model or apply a nonparametric test for each feature independently, usually following normalization and a logarithmic or log-ratio-based transformation and ensuring that the model and method reflect the study design. The R package MaAsLin3 can automate this process, including the prior application of normalization, transformations, filtering, and subsequent multiplicity corrections, while incorporating both fixed and random covariates in linear models.

However, more sophisticated analysis approaches have been proposed to identify differentially abundant microbiome features while better addressing the specific challenges associated with microbial data.

Widely used methods and their broad approaches are outlined below. A thorough methodological review of methods for microbiome data analysis including differential abundance methods is given by Petersen et al. [19], although the field is rapidly developing and up-to-date guidance on the theory and performance of different methods should always be sought:

1. ANCOM-BC2 (Analysis of Composition of Microbiomes with Bias Correction) [20] estimates differential abundance using log-linear models incorporating covariates and repeated measures. ANCOM-BC2 includes a bias correction to account for sampling variation and uneven sequencing depth, specific methods to control the high variance associated with small counts and zeros, and a modified multiple testing correction procedure to account for multi-group testing.
2. ALDEx2 (ANOVA-Like Differential Expression) [21] relies on resampling to model the uncertainty associated with small counts and zeros, followed by application of log-ratio transformations and pooling of the subsequent statistical testing and effect size estimation across resampled replicates.
3. MaAsLin3 (Multivariate Association with Linear Models) specializes in identifying associations between microbial features and multiple covariates. The linear modeling framework used by the tool allows for complex study designs, including multiple fixed and random covariates, and enables absolute abundance comparisons through the incorporation of qPCR data or spike-ins [22].
4. LEfSe (Linear discriminant analysis Effect Size) is an older method that combines nonparametric statistical significance

testing with linear discriminant analysis to identify features most likely to explain differences between conditions [23]. However, LEfSe does not include a multiple testing correction, and so is prone to false positives unless a separate correction is applied to results, it cannot accommodate covariates, and provides effect sizes that are difficult to interpret, particularly with more than two groups.

5. DESeq2, edgeR, and limma have been used for differential abundance analysis but were originally developed for RNA transcriptomic experiments. They benefit from well-developed software packages and wide acceptance in the field. However, they rely on distributional assumptions that may be questionable in microbiome data, so they should be used with caution [10].

2.4.4 Selecting Methods and Models

There are many methodological decisions in microbiome preprocessing and statistical modeling steps, not only in the selection of which methods to use but in how they are applied. Keep in mind the principles outlined in Subheading 1 when selecting which methods to apply.

Prior research can provide a useful guide to the possible visualization or analytical approaches that might be used, but published analyses are not always suitable, and analyses that are appropriate for one situation may not be appropriate in another.

The choice between methods can be informed by study design, research questions, and empirical comparisons of the robustness of different methods. Several useful reviews of methods for the analysis of microbiome data exist [24–28] as well as comparisons of their validity using real and simulated data that can help to evaluate their suitability for a particular design. In 2022, Nearing et al. [29] found a considerable difference in the number of differential taxa identified and in which taxa were identified as differential between groups of samples in a comprehensive test of different but commonly used analysis methods combined with different normalization and filtering procedures applied to real datasets. Although it is impossible to know which method produced the "correct" result when tested with real datasets, some methods appear to perform consistently less well across a range of scenarios than others.

- With simple designs, ANCOM-BC or ALDEX appear the most consistent and conservative. Other methods including LefSE and edgeR report more differentially abundant taxa but are likely to include more false positives and should be avoided [29].
- For more complex study designs, methods based on straightforward repeated application of linear models such as MaAsLin can be used with care, ensuring that the linear models being

applied are appropriate, and the selection of transformations to apply and models within these methods are carefully considered.

- For designs that cannot be accommodated by these specific methods, it may be necessary to estimate linear models using traditional software or use other standard statistical approaches, again being careful that appropriate normalization, transformation, and multiple testing correction have been applied to deal with the specific characteristics of the metagenomic dataset. Sensitivity analyses comparing results across multiple methods can provide greater confidence in identified differentially abundant features.
- Optimal normalization methods may depend on study goals. For example, while log-ratios provide a theoretical way to overcome the problems of compositionality, simulations have shown that simpler normalization and transformation may, in practice, enable better results for classification problems [30].

Also, consider carefully which covariates are important to include in the analysis. Confounding factors that are not included in the analysis might introduce false positive relationships between the primary exposure and outcome of interest. Incorporating covariates can also help reduce unexplained variance in the data, potentially increasing the power to detect true associations.

Potential covariates are discussed in Chapter 2 but might include patient-level characteristics like age and sex, technical factors such as sequencing batch, experimental blocks, or library size, or baseline values of outcomes in the case of longitudinal data analysis.

2.4.5 Predictive Models and Machine Learning

If the goal of an analysis is to develop and validate the microbiome as a predictor (as opposed to understanding the microbiome as a cause or consequence of another factor), then the considerations and methods differ from the analyses discussed above.

It is less important in this case to identify the specific taxa that are causally linked to an outcome, or to demonstrate statistically significant effects or differences between groups. What is important is the accuracy with which a model based on the microbial data (with or without other aspects of metadata) can predict an outcome, and whether this can be generalized to a wider population. Machine learning methods can be used to develop predictive models, which should be internally cross-validated within a cohort or externally validated with a test cohort that is separate from the cohort in which the model was developed. Machine learning models are reviewed by Marcos-Zambrano [31].

2.4.6 Analysis of Gene Abundance or Functional Pathways

As well as differences in the presence or abundance of microbial taxa, interest is often in comparing gene-specific or functional composition of the microbiome. This can proceed using similar methods to those used for the analysis of microbial taxon abundances.

As with testing for differential taxon abundances, a high number of features means that false positives are a concern. To mitigate this, functional hypotheses need to be as targeted as possible. One way to reduce the feature space that is posed by cataloging single enzymes is to agglomerate these to higher functional groups, e.g., metabolic pathways that depend on a set of enzymes. This can be conducted by, for example, summing the abundance of enzymes within a pathway (in the simplest case), investigating for functional enrichments such as gene set enrichment analysis, or using explicit genome-scale metabolic modeling. These pathway matrices summarized at higher functional levels can then be investigated via multivariate or univariate statistical methods described for taxon abundances, including multiple testing correction.

2.4.7 Identifying the Experimental Unit and Avoiding Pseudo-Replication

Many experimental designs can induce shared variance between samples. For example, repeated sampling from the same individual, within households, or sibling pairs will produce samples that may be more like each other than samples from different patients or households. In nonhuman studies, samples from mice housed in the same cage might be more like each other than samples from mice housed in different cages.

Simple non-parametric tests, ordinary linear models, PERMANOVA, and ANOVA assume by default that samples are independent when calculating standard errors and *p*-values. These methods can have extremely high false-positive rates if these potential correlations are ignored, which is often the case in published literature [32].

1. Identify the potential sources of shared variance between samples in your dataset.
2. Use a repeated-measures ANOVA or a linear mixed model, or methods such as ANCOM-BC2, or Maaslin [33] that can incorporate random effects, refer to the documentation to ensure that the correct terms are included.
3. Do not use methods that cannot accommodate important covariates or shared variance in your experimental design.
4. When applying a permutation test such as PERMANOVA, make sure that your permutations are correctly restricted such that each permutation is valid under your experimental design.
5. If this is impossible, depending on the design and the research question, it might be acceptable to “average” samples within

clusters (e.g., by pooling repeated measures) and then treating each *cluster* as an independent observation.

2.4.8 Correcting for Multiplicity

When many null hypotheses are tested, the chance that one or more are rejected even though they are true increases, leading to a high rate of potential false-positive results. There are two traditional approaches to prevent this: control for the *false discovery rate* (FDR) or control for the *family-wide error rate* (FWER). FDR corrections are done automatically by many software packages, so when interpreting results, be aware whether corrected or uncorrected *p*-values are being reported.

FDR control (e.g., Benjamini-Hochberg or *q*-value correction) is usually applied to exploratory analyses with many outcomes where a certain proportion of false positives can be tolerated. FWER control (e.g., Bonferroni corrections) is more typically applied to small numbers of more specific hypotheses, to control the possibility of *any* false positives arising.

When an FDR correction is applied to *p*-values, their meaning changes. An "uncorrected" *p*-value is the probability that an association at least as large as the one observed would occur, even if the null hypothesis (typically of no association) was true. When an FDR correction is applied and a threshold determined, this value is the expected proportion of "significant" results (those below the threshold) that are false positives. False discovery rates of 0.05 or 0.1 are typically accepted; these should be specified in reports.

Fortunately, if no or very few true relationships exist in your study, then controlling the FDR at 5% also controls the FWER at close to 5%, that is, the probability of drawing one or more false-positive conclusions is also 5%.

1. When more than one hypothesis is being tested, then multiplicity correction to *p*-values should be made. Always use a multiplicity correction when conducting an exploratory analysis of many outcomes.
2. Use FDR corrections (such as Benjamini-Hochberg or *q*-value corrections) because they are standard in the field but be sure their interpretation is understood.
3. Always describe multiplicity adjustments fully in manuscripts and decide on these in advance as part of your analysis plan.
4. Choose your FDR threshold before starting the analysis. Setting an FDR threshold of 0.05 is good practice, as, in the case that all hypotheses are false, this will control the FWER (the overall chance you make a false positive at 5%). However, the specific aims might require a different threshold, depending on the tolerance for false positives.

5. In any case, keep in mind that *p*-values and corrected *p*-values are measures of the strength of evidence and do not determine with certainty whether associations are present or not.

2.4.9 Presenting Results from Differential Abundance Analyses

Differential abundance methods will usually return lists of features along with their relative abundance across conditions, estimates of "effects" such as fold-changes in abundance, and measures of whether these effects are statistically significant. Effects and significant features can be visualized using different methods. Volcano plots can be used to display both statistical significance and effect size, with significant features highlighted. Heatmaps can visualize differential abundance patterns across multiple features and samples simultaneously. When reporting results, it is important to include multiplicity-adjusted *p*-values and the chosen threshold for statistical significance. Effect sizes should be reported alongside statistical significance to improve interpretation. For taxonomic features, users can explore the possibility to display their results within the context of the phylogenetic tree to identify patterns across related taxa, using packages such as Metacoder for R [34].

2.5 Model Validation, Diagnostics, and Sensitivity Analysis

Be aware of the assumptions underlying statistical tests being used and the available methods to assess these. Model validation ensures the robustness and reliability of the model's conclusions. This involves checking a model's assumptions and sensitivity analysis.

Verifying the homoscedasticity and normality assumptions through basic diagnostics plot (e.g., Q-Q plots or residual plots) is important before interpreting the results of methods that rely on linear regression. For more complex models such as mixed-effects models, model convergence should be checked, and warnings from estimation software should not be ignored. Non-convergence of a model may suggest insufficient data to estimate all parameters of the model, or a model that does not fit the dataset, in which case the results may not be valid, and the model must be respecified.

When estimating many models simultaneously, either with a custom script or using a specific package such as Maaslin3, it is difficult to examine every model that is estimated. In this case, plotting the raw and normalized data along with validation checks for those models that lead to key findings should be performed, to ensure that they are not driven by outliers or poor modeling choices. Some packages including deseq2 offer diagnostic plots to check that the distribution of the data across features conforms to the modeling assumptions, and these should be used where available.

To assess the false positive rate of a modeling approach, methods can be tested on real datasets that have been modified to eliminate genuine relationships. For example, researchers can select a homogeneous subset of data, randomly partition it into groups, and then test for differences between those groups. If large numbers of differences are still consistently detected, then the method is unlikely to be suitable.

None of the current approaches to microbiome analysis is without limitations, and different methods can lead to different results or conclusions. Sensitivity analyses can be used to test to what extent findings are dependent on choices of normalization methods, transformations, covariate selection, missing data modeling, exclusion of outlying data, quality control thresholds, or modeling approaches. These should be pre-planned as far as possible, and any substantive differences in findings with sensitivity analyses should be reported alongside the main findings.

3 Software Considerations

Some microbiome analyses can be performed with standard statistical routines and software. However, microbiome datasets are complex, and many common data management, preprocessing, and analysis tasks along with specifically designed data structures are implemented in R packages such as PhyloSeq [35], Mia [36], and vegan [37] and in Python libraries including scikit-bio [38].

Specific methods developed for, for example, differential abundance analysis are usually implemented in specific R packages available through the Bioconductor repository and interact well with these data structures. Many are available in QIIME 2 [39] or Galaxy workflows or have been implemented as web-based interfaces such as MicrobiomeAnalyst [40].

The choice of platform depends on the specific analytical needs and user expertise. Standard analyses based on simple experimental designs are more likely to be manageable using web-based solutions or graphical interfaces, but for more complex designs, custom scripts may be needed. In all cases, it is important that the analyst understands the methods being used, their suitability for the dataset and the experimental design, the options being presented, and takes responsibility for their suitability for the task.

4 General Tips for the Statistical Analysis of Microbiome Data

4.1 Understand Your Analysis and Its Limitations

It is important to understand the broad approach taken by the analytical methods being used, their assumptions, limitations, and how to interpret their outputs. The software documentation, academic papers, tutorials, and vignettes that typically accompany computational methods are important sources of guidance for their correct application, interpretation, and incorporation within workflows and should be studied before use.

4.2 Accept Negative and Equivocal Findings, Interpret p-Values Correctly

- Avoid overinterpreting p-values as definitive evidence of biological relevance. A small p-value indicates statistical evidence against the null hypothesis, but does not measure effect size, clinical importance, or reproducibility. Especially in high-dimensional microbiome data, nominal significance can arise by chance, particularly if statistical models do not fit data well as can be the case with highly over-dispersed count data.
- Conversely, remember that "absence of evidence is not evidence of absence" and that lack of a significant p-value alone is not sufficient to conclude that no relationship exists, particularly where sample sizes are small, and power is low.
- Always quantify uncertainty in a finding honestly, particularly for small exploratory studies where estimates will be imprecise, and power is low. Conclusive results (e.g., whether a particular microbial feature is or is not associated with a covariate) may not be possible.
- Emphasize effect sizes, confidence intervals, and consistency across methods or datasets alongside p-values to support robust and interpretable findings. This will communicate the size (and likely importance) of an effect, as well as the uncertainty surrounding it.
- Keep in mind that it is possible that the microbiome has no relationship with the exposure, treatment, or outcome that you are investigating, and you should consider that demonstrating a lack of effect can be as important a contribution to the scientific literature, as demonstrating associations.

4.3 Ensure Workflows Are Reproducible

- Ensure that all stages of statistical analyses are recorded and documented, including version numbers of software packages used.
- Where permutation tests or other stochastic procedures are used, make sure that the random seed is recorded.
- If using a graphical user interface for analysis, as opposed to writing code, make sure you can record exactly what steps you have taken.

4.4 Be Aware of the Wider Statistical Issues Surrounding Your Analysis

All statistical analyses have potential pitfalls that are not related to the "microbiome" context. These are less widely known than they should be in the microbiology community and so lead to many inefficient or invalid analyses being produced and published.

- Study the appropriate general statistical literature related to your study design (whether in vitro, preclinical, clinical trial, or epidemiological study).

- For example, avoid dangers associated with regression to the mean [41], inappropriate use of baseline data [42], and incorrect interpretation of subgroup analyses.
- For an epidemiologic study, consider the risk of spurious correlations from omitted confounders and how selection bias might influence your associations [43]. For case-control studies, ensure that matching factors or indicators are included as covariates [44].

4.5 Selecting Covariates to Include in Analysis

Any factor that might influence an outcome variable should be considered as a possible covariate to be included in the statistical analysis.

Systematic biases and between-sample variance can also be introduced during the processing and sequencing of samples, and so these should be understood, and relevant factors incorporated into analyses where possible.

4.6 Avoid Questionable Research Practices

The large number of analytical choices that need to be made before analysis of a microbiome dataset means that it is easy to unknowingly commit "*questionable research practices*" (QRPs) [45]. QRPs include any practice employed to "spin" research findings, either intentionally or unintentionally. These include (but are not limited to):

- Selecting methods, models, or subgroups because they lead to positive results, ignoring negative findings or findings that do not fit the story that you want to tell.
- Deciding on the hypotheses to be tested after the results are known (HARKing).
- Failing to correct for multiplicity to preserve statistical significance.
- Ignoring known limitations of methods.
- Presenting the results of an exploratory experiment as if they are confirmatory, presenting a secondary or sensitivity analysis as if it was the primary analysis.
- Overstating the confidence in results or the importance of findings.

QRPs can dramatically inflate false-positive rates even though researchers present analytic decisions that would, in isolation, seem reasonable. Microbiome research is particularly susceptible to QRPs because: (1) Findings often vary enormously with methods; (2) many methods have high false positive rates; and (3) many different hypotheses are typically tested in a single analysis of microbiome data. To avoid QRPs leading to false-positive findings:

- Pre-specify and register statistical analysis plans as far as possible and document the reasons for deviations from the original analysis plan.
- Prevent unconscious bias by ensuring analysts are blinded to group allocations, particularly where a result in a particular direction is hypothesized.
- Accept negative results as being likely, valid, and important research findings.
- Avoid HARKing (hypothesizing after the results are known). Be explicit and honest when a statistical comparison has been motivated by an association noticed within the dataset.
- Use recognized reporting guidelines for academic papers (from the EQUATOR network [46]) and for microbiome-specific reports [47] and ensure the significance or reliability of your conclusions is not overstated.

5 Conclusions

Successful analysis of microbiomes requires an understanding of good statistical analysis practice in general, and the additional complications associated with measuring microbiomes and the complexity of the questions that might be asked of them. Datasets are typically sparse, high-dimensional compositional count data with many more features than samples. Many methods have been developed and are currently under development to perform statistical analysis of microbiome data. Any reasonable method can be used for data analysis, but none are without significant limitations. Analysts should critically evaluate methods and results with respect to their own dataset, as they should for any statistical analysis.

Acknowledgements

This work was funded by the Biotechnology and Biological Sciences Research Council (BBSRC) through an Institute Strategic Programme award to the QIB programme Food, Microbiome and Health BB/X011054/1 and its constituent projects, and the Core Capability Grants BB/CCG1860/1 and BB/CCG2260/1.

References

1. Corander J, Hanage WP, Pensar J (2022) Causal discovery for the microbiome. Lancet Microbe 3:e881–e887. https://doi.org/10.1016/S2666-5247(22)00186-0
2. Lipsky AM, Greenland S (2022) Causal directed acyclic graphs. JAMA 327:1083–1084. https://doi.org/10.1001/jama.2022.1816

3. Yuan I, Topjian AA, Kurth CD et al (2019) Guide to the statistical analysis plan. Pediatr Anesth 29:237–242. https://doi.org/10.1111/pan.13576
4. Tsilimigras MCB, Fodor AA (2016) Compositional data analysis of the microbiome: fundamentals, tools, and challenges. Ann Epidemiol 26:330–335. https://doi.org/10.1016/j.annepidem.2016.03.002
5. Gloor GB, Macklaim JM, Pawlowsky-Glahn V, Egozcue JJ (2017) Microbiome datasets are compositional: and this is not optional. Front Microbiol 8:2224. https://doi.org/10.3389/fmicb.2017.02224
6. Xia Y (2023) Statistical normalization methods in microbiome data with application to microbiome cancer research. Gut Microbes 15(2): 2244139. https://doi.org/10.1080/19490976.2023.2244139
7. Schloss PD (2024) Rarefaction is currently the best approach to control for uneven sequencing effort in amplicon sequence analyses. mSphere 9:e00354-23. https://doi.org/10.1128/msphere.00354-23
8. Paulson JN, Stine OC, Bravo HC, Pop M (2013) Robust methods for differential abundance analysis in marker gene surveys. Nat Methods 10:1200–1202. https://doi.org/10.1038/nmeth.2658
9. Hong J, Karaoz U, de Valpine P, Fithian W (2022) To rarefy or not to rarefy: robustness and efficiency trade-offs of rarefying microbiome data. Bioinformatics 38:2389–2396. https://doi.org/10.1093/bioinformatics/btac127
10. Boshuizen HC, te Beest DE (2023) Pitfalls in the statistical analysis of microbiome amplicon sequencing data. Mol Ecol Resour 23:539–548. https://doi.org/10.1111/1755-0998.13730
11. Silverman JD, Washburne AD, Mukherjee S, David LA (2017) A phylogenetic transform enhances analysis of compositional microbiota data. elife 6:e21887. https://doi.org/10.7554/eLife.21887
12. Hülpüsch C, Rauer L, Nussbaumer T et al (2023) Benchmarking MicrobIEM – a user-friendly tool for decontamination of microbiome sequencing data. BMC Biol 21:269. https://doi.org/10.1186/s12915-023-01737-5
13. Brüssow H (2020) Problems with the concept of gut microbiota dysbiosis. Microb Biotechnol 13:423–434. https://doi.org/10.1111/1751-7915.13479
14. Bray JR, Curtis JT (1957) An ordination of the upland Forest communities of southern Wisconsin. Ecol Monogr 27:325–349. https://doi.org/10.2307/1942268
15. Lozupone CA, Hamady M, Kelley ST, Knight R (2007) Quantitative and qualitative β diversity measures lead to different insights into factors that structure microbial communities. Appl Environ Microbiol 73:1576–1585. https://doi.org/10.1128/AEM.01996-06
16. Legendre P, Anderson MJ (1999) Distance-based redundancy analysis: testing multispecies responses in multifactorial ecological experiments. Ecol Monogr 69:1–24
17. Anderson MJ (2017) Permutational Multivariate Analysis of Variance (PERMANOVA). In Wiley StatsRef: Statistics Reference Online (eds N. Balakrishnan, T. Colton, B. Everitt, W. Piegorsch, F. Ruggeri and J.L. Teugels). https://doi.org/10.1002/9781118445112.stat07841
18. Simpson (2025) Restricted permutations; using the permute package (https://cran.r-project.org/web/packages/permute/vignettes/permutations.html)
19. Peterson CB, Saha S, Do K-A (2024) Analysis of microbiome data. Annu Rev Stat Its Appl 11:483–504. https://doi.org/10.1146/annurev-statistics-040522-120734
20. Lin H, Peddada SD (2024) Multigroup analysis of compositions of microbiomes with covariate adjustments and repeated measures. Nat Methods 21:83–91. https://doi.org/10.1038/s41592-023-02092-7
21. Fernandes AD, Reid JN, Macklaim JM et al (2014) Unifying the analysis of high-throughput sequencing datasets: characterizing RNA-seq, 16S rRNA gene sequencing and selective growth experiments by compositional data analysis. Microbiome 2:15. https://doi.org/10.1186/2049-2618-2-15
22. Nickols WA, Kuntz T, Shen J, et al (2024) MaAsLin 3: refining and extending generalized multivariable linear models for meta-omic association discovery. bioRxiv 2024.12.13.628459. https://doi.org/10.1101/2024.12.13.628459
23. Segata N, Izard J, Waldron L et al (2011) Metagenomic biomarker discovery and explanation. Genome Biol 12:R60. https://doi.org/10.1186/gb-2011-12-6-r60
24. Knight R, Vrbanac A, Taylor BC et al (2018) Best practices for analysing microbiomes. Nat Rev Microbiol 16:410–422. https://doi.org/10.1038/s41579-018-0029-9
25. Kleine Bardenhorst S, Berger T, Klawonn F et al (2021) Data analysis strategies for microbiome studies in human populations-a systematic review of current practice. mSystems

6. https://doi.org/10.1128/mSystems.01154-20

26. Paliy O, Shankar V (2016) Application of multivariate statistical techniques in microbial ecology. Mol Ecol 25:1032–1057. https://doi.org/10.1111/mec.13536
27. McMurdie PJ (2018) Normalization of microbiome profiling data. In: Beiko RG, Hsiao W, Parkinson J (eds) Microbiome analysis: methods and protocols. Springer, New York, pp 143–168
28. Weiss S, Xu ZZ, Peddada S et al (2017) Normalization and microbial differential abundance strategies depend upon data characteristics. Microbiome 5:27. https://doi.org/10.1186/s40168-017-0237-y
29. Nearing JT, Douglas GM, Hayes MG et al (2022) Microbiome differential abundance methods produce different results across 38 datasets. Nat Commun 13:342. https://doi.org/10.1038/s41467-022-28034-z
30. Yerke A, Fry Brumit D, Fodor AA (2024) Proportion-based normalizations outperform compositional data transformations in machine learning applications. Microbiome 12:45. https://doi.org/10.1186/s40168-023-01747-z
31. Marcos-Zambrano LJ, Karaduzovic-Hadziabdic K, Loncar Turukalo T et al (2021) Applications of machine learning in human microbiome studies: a review on feature selection, biomarker identification, disease prediction and treatment. Front Microbiol 12:634511. https://doi.org/10.3389/fmicb.2021.634511
32. McKinnon Reish H, Dewey L, Kirschman LJ (2024) A host of issues: pseudoreplication in host–microbiota studies. Appl Environ Microbiol 90:e01033–e01024. https://doi.org/10.1128/aem.01033-24
33. Mallick H, Rahnavard A, McIver LJ et al (2021) Multivariable association discovery in population-scale meta-omics studies. PLoS Comput Biol 17:e1009442. https://doi.org/10.1371/journal.pcbi.1009442
34. Foster ZSL, Sharpton TJ, Grünwald NJ (2017) Metacoder: an R package for visualization and manipulation of community taxonomic diversity data. PLoS Comput Biol 13:e1005404. https://doi.org/10.1371/journal.pcbi.1005404
35. McMurdie PJ, Holmes S (2013) Phyloseq: an R package for reproducible interactive analysis and graphics of microbiome census data. PLoS One 8:e61217. https://doi.org/10.1371/journal.pone.0061217
36. Borman T, Ernst F, Lahti L (2025). miaViz: Microbiome Analysis Plotting and Visualization. https://doi.org/10.18129/B9.bioc.miaViz, R package version 1.16.0, https://bioconductor.org/packages/miaViz.
37. Dixon P (2003) VEGAN, a package of R functions for community ecology. J Veg Sci 14:927–930. https://doi.org/10.1111/j.1654-1103.2003.tb02228.x
38. Rideout JR, Caporaso G, Bolyen E et al (2025) scikit-bio/scikit-bio: scikit-bio 0.6.3. https://doi.org/10.5281/zenodo.14640761
39. Kuczynski J, Stombaugh J, Walters WA et al (2012) Using QIIME to analyze 16S rRNA gene sequences from microbial communities. Curr Protoc Microbiol Chapter 1: Eunit 1E.5. https://doi.org/10.1002/9780471729259.mc01e05s27
40. Lu Y, Zhou G, Ewald J et al (2023) MicrobiomeAnalyst 2.0: comprehensive statistical, functional and integrative analysis of microbiome data. Nucleic Acids Res 51:W310–W318. https://doi.org/10.1093/nar/gkad407
41. Bland JM, Altman DG (1994) Statistics notes: some examples of regression towards the mean. BMJ 309:780. https://doi.org/10.1136/bmj.309.6957.780
42. Vickers AJ, Altman DG (2001) Analysing controlled trials with baseline and follow up measurements. BMJ 323:1123–1124. https://doi.org/10.1136/bmj.323.7321.1123
43. Westreich D (2012) Berkson's bias, selection bias, and missing data. Epidemiol Camb Mass 23:159–164. https://doi.org/10.1097/EDE.0b013e31823b6296
44. Pearce N (2016) Analysis of matched case-control studies. BMJ 352:i969 https://doi.org/10.1136/bmj.i969
45. Fraser H, Parker T, Nakagawa S et al (2018) Questionable research practices in ecology and evolution. PLoS One 13:e0200303. https://doi.org/10.1371/journal.pone.0200303
46. The EQUATOR Network | Enhancing the QUAlity and Transparency Of Health Research. https://www.equator-network.org/. Accessed 13 May 2021
47. Mirzayi C, Renson A, Zohra F et al (2021) Reporting guidelines for human microbiome research: the STORMS checklist. Nat Med 27:1885–1892. https://doi.org/10.1038/s41591-021-01552-x

Chapter 13

Profiling Microbial Metabolites in Human Urine, Fecal, and In Vitro Colon Model Samples

Jennifer Ahn-Jarvis, Marianne Defernez, Gwénaëlle Le Gall, and Michael D. Paxhia

Abstract

The human gut microbiome is a complex ecosystem that contributes to the health of the host by transforming nutrients from the diet. Many of these interactions are mediated through the exchange of metabolites between the microbes and the host. Here, we describe two complementary targeted metabolomics techniques, liquid chromatography-mass spectrometry (LC-MS) and nuclear magnetic resonance (NMR), to monitor specific changes in metabolite abundance in samples from human urine, fecal, and in vitro GI colon model systems. This approach offers the specificity and sensitivity of LC-MS quantification with an understanding of broad changes in metabolite abundance by NMR that together can aid in understanding how the gut microbiome metabolizes different dietary nutrients.

Key words SCFA, BCFA, Targeted metabolomics, Microbiome, LC-MS, NMR

1 Introduction

The human gut microbiome encodes a diversity of metabolic pathways involved in transforming metabolites from food into bioactive molecules that can affect the host [1, 2]. One major class of metabolites is fatty acids, including short-chain and branched-chain fatty acids (SCFAs and BCFAs). SCFAs are associated with the gut microbiome fermenting sugars derived from dietary fiber and resistant starches and include acetate, propionate, butyrate, and valerate [3, 4]. BCFAs originate from the fermentation of branched-chain amino acids and include isobutyrate, isovalerate, and 2-methylbutyrate [3]. SCFAs have wide-ranging physiological effects including reducing intestinal inflammation, improving host glucose homeostasis, inducing serotonin production, and limiting the progression of colon cancer [3–7]. Here, we present techniques for the collection and targeted metabolomic analysis of urine (*see* **Note 1**), feces, and samples from an exemplar in vitro fermentation system, a colon model (*see* **Note 2**), by LC-MS and NMR (*see* **Note**

Simon R. Carding (ed.), *Best Practice in Microbiome Research*, Springer Protocols Handbooks,
https://doi.org/10.1007/978-1-0716-5009-7_13,

3) with a focus on SCFAs and BCFAs as model gut microbiome-derived metabolites. Multivariate analysis can be used to investigate associations between samples, as well as highlight clustering between multiple metabolites that change in concert with one another across samples and hence can be key to understanding metabolic signatures of different experimental treatments. The principles behind these techniques can be adapted and applied to better understand the impact of the microbiome on metabolism of the host in in vitro, mouse, and human dietary intervention studies. Two complementary techniques used for targeted metabolomics include LC-MS and NMR to gain a comprehensive understanding of biological systems. Each approach has its advantages and disadvantages, described in Table 13.1.

2 Materials

2.1 General Materials

1. Vortex.
2. Microcentrifuge tubes.
3. Micropipettes and aerosol-resistant pipette tips.

1. Reusable cooler box 4 °C for 96 hours (DSX 12, Diagnosach; Tempack®, Barcelona, ESP).

Table 13.1
Comparison of attributes of LC-MS and NMR

Attributes	LC-MS [8]	NMR [9]
Sensitivity	pM to nM	µM to mM
Quantification	Absolute quantification. Capable of quantifying 200 to 400 metabolites	Absolute quantification. Capable of quantifying 50 to 100 metabolites
Selectivity	Distinguish isomers	Poorly detects isomers or trace compounds
Sample preparation	Complex, often requiring derivation, destruction of sample	Minimal, direct analysis of samples, sample can be recovered
Matrix effect	Considerable matrix effect and requires careful method optimization relying on isotopic internal standards	Minimal matrix effects
Structural identification	Structural identification limited, requires tandem MS	Structural identification of unknown compounds
Reproducibility	Reproducibility affected by experimental conditions	Highly reproducible

2.2 Urine Collection and Preparation for Analysis

2.2.1 Essential Equipment

2. Balance (4.2 kg capacity, Explorer Ohaus™, Parsippany, USA).
3. −80 °C Freezer.

2.2.2 Consumables

1. Ice packs (FCP FF600, Flexible cold pack; Tempack®, Barcelona, ESP).
2. Pre-weighed and labeled 24 h urine collection container (Uri-safe® 3 L, B350; Simport®, Beloeil, CAN).
3. Urine collection facilitator (CMS-106, Caremax Supply, Buena Park, USA).
4. Gloves and hand sanitizer upon participant request.
5. Resealable (Ziploc) plastic bags (2) and absorbent pad (compliance to Packing Instruction 650 for transport of biologically hazardous materials, UN3373).
6. Urinalysis Test Strips (Multistix® 10 SG; Siemens, Munich, DEU).
7. 1-mL footed cryogenic tubes (Nunc™, Thermo Scientific™, Waltham, USA)
8. Pipette, graduated 1 ml sterile, single wrapped Pastette®.

2.3 Serum Bottle *In Vitro* Colon Model

2.3.1 Essential Equipment

1. Anaerobic cabinet (e.g., Whitely A95 anaerobic workstation).
2. Class 2 Microbiological Safety Cabinet.
3. Stomacher 400 Circulator Lab blender.
4. Stomacher 400 classic bags.
5. Wheaton glass serum bottles, 125 mL.
6. Wheaton closed-top seals.
7. Grey-butyl rubber stoppers, 20 mm diameter.
8. Hand-operated aluminium cap crimper, 20 mm diameter.
9. Vial Decapper, 20 mm diameter.
10. 19-gauge luer-lock needles
11. 1 mL, 5 mL, 10 mL syringes
12. MP Bio FastDNA Spin Kit for Soil.

2.3.2 Stock Solutions and Reagents

1. Basal medium—Trypticase Peptone, NaCl, KCl, $CaCl_2{\bullet}2H_2O$, $MgSO_4{\bullet}7H_2O$, PIPES, NH_4Cl, and KOH.
2. Hemin solution—Hemin, NaOH.
3. Resazurin solution—Resazurin.

4. Fatty acid solution—Acetic acid, propionic acid, butyric acid, isobutyric acid, 2-methylbutyric acid, valeric acid, isovaleric acid, and NaOH.
5. Trace mineral solution—$MnCl_2{\bullet}4H_2O$, $FeSO_4{\bullet}7H_2O$, ZnCl, $CuCl{\bullet}2H_2O$, $CoCl_2{\bullet}6H_2O$, SeO_2, $NiCl_2{\bullet}6H_2O$, $Na_2MoO_4{\bullet}2H_2O$, $NaVO_3$, H_3BO_3, and HCl.
6. Vitamin and phosphate solution—KH_2PO_4, biotin, folic acid, DL-pantothenic acid hemicalcium salt, nicotinamide, riboflavin, thiamine HCl, pyridoxine HCl, 4-aminobenzoic acid, and cyanocobalamin.
7. Sodium carbonate solution—Na_2CO_3.
8. Reducing agent solution—L-cysteine HCl, $Na_2S{\bullet}9H_2O$, and NaOH.
9. Fermentable carbohydrate of interest—e.g., arabinogalactan (larch wood) or pectin (apple).

2.4 Targeted Quantification of SCFA and BCFA by LC-MS/MS

2.4.1 Essential Equipment

1. Agilent 6490 Triple Quad MS mass spectrometer.
2. Agilent 1290 HPLC system with autosampler.
3. Thermo Scientific Hypercarb® column (PGC, 3 μm, 50 mm × 2.1 mm) with guard column.

2.4.2 Reagents

1. Fatty acids: acetic acid, butyric acid, isobutyric acid, isovaleric acid, propionic acid, valeric acid.
2. Isotopically labeled fatty acid standards: D_4-acetic acid, $^{13}C_2$-butyric acid, D_6-isobutyric acid, D_9-isovaleric acid, D_2-propionic acid, D_9-valeric acid.
3. Mobile phase materials: LC-MS-grade acetonitrile, Milli-Q or LC-MS-grade water, Suprapur® grade formic acid.
4. Autosampler vials and caps.

2.5 NMR Metabolomics

2.5.1 Essential Equipment

1. Bruker Avance Neo 600 MHz NMR spectrometer with cryoprobe and a 60-position autosampler.
2. 5-mm 7″ Norell NMR tubes (*see* **Note 4**).

2.5.2 Reagents

1. Solvents: 99.9% deuterium oxide (D_2O), MilliQ water, or equivalent.
2. Phosphate buffer materials: NaH_2PO_4, K_2HPO_4, NaN_3.
3. Reference standard: 3-(trimethylsilyl) propionic-2,2,3,3-d4 acid sodium salt (TSP).

3 Methods

3.1 Urine Collection

Participants are provided step-by-step instructions on how to use the collection vessel and as needed a collection facilitator for women participants to aid in their urine collection (*see* **Note 5**). Participants are instructed to collect all their urine for a 24-hour period and return the collection vessel back in the cool box since no preservatives are used (*see* **Note 6**).

1. Upon receiving the completed 24-hour urine collection, weigh 24 h urine collection and subtract the mass of the collection vessel to determine the mass of the urine.
2. Collect one urine testing strip, and using the Pasteur pipette, collect 1 mL of the urine sample and use it to wet the entire test strip.
3. After 60 s (follow manufacturer instructions since these times may vary amongst different manufacturers), read the results and record the specific gravity, pH, and if there are any incidental findings that may interfere with the metabolomic analysis such as blood and/or leukocytes (*see* **Note 7**).
4. Aliquot 1 mL urine using a Pasteur pipette from the urine collection container into 1 mL cryogenic storage tubes and store at −80 °C until analysis (*see* **Note 8**).

3.2 Serum Bottle *In Vitro* Colon Model

Fecal inoculum is a standardized mixture prepared from human stool samples, used to inoculate in vitro colon models for studying gut microbiota and fermentation processes (*see* **Note 9**), see Chapter 3. Fecal extracts can be prepared directly and analyzed by NMR if anaerobic culturing is unavailable (see Subheading 3.4.3 and Chapter 3 for relevant information).

3.2.1 Stock Solutions

1. Hemin solution—Dissolve 0.05 g hemin in 25 mL 0.05 M NaOH, bring to a 500 mL total volume with Milli-Q H_2O and filter sterilize. Cover the bottle with foil and store at 4 °C in the dark.
2. Trace mineral solution—Dissolve 0.86 mL concentrated 12 N HCl in 400 mL ddH_2O. Add the amounts of trace minerals in order, letting each fully dissolve before adding the next: 12.5 mg $MnCl_2{\bullet}4H_2O$, 10 mg $FeSO_4{\bullet}7H_2O$, 12.5 mg ZnCl, 12.5 mg $CuCl{\bullet}2H_2O$, 25 mg $CoCl_2{\bullet}6H_2O$, 25 mg SeO_2, 125 mg $NiCl_2{\bullet}6H_2O$, 125 mg $Na_2MoO_4{\bullet}2H_2O$, 15.7 mg $NaVO_3$, and 125 mg H_3BO_3. Bring the total volume up to 500 mL with Milli-Q H_2O, cover with foil, and autoclave.
3. Resazurin solution—Dissolve 0.05 g resazurin in 50 mL Milli-Q H_2O, filter sterilize, and store covered with foil at 4 °C.

4. Fatty acid solution—Add 0.8 g NaOH to 80 mL ddH_2O and add 685 μL acetic acid, 300 μL propionic acid, 184 μL butyric acid, 47 μL isobutyric acid, 55 μL 2-methylbutyric acid, 55 μL valeric acid, and 55 μL isovaleric acid. Bring to a final volume of 100 mL with Milli-Q H_2O and store at 4 °C.
5. Sodium carbonate solution—Dissolve 41 g of sodium carbonate into boiled ddH_2O. Bring to a final volume of 500 mL with boiled ddH_2O and bubble with CO_2 for 30 min. Aliquot 100 mL into serum bottles, cap, and autoclave.
6. Vitamin and phosphate solution—Dissolve 27.35 g KH_2PO4 in 400 mL ddH_2O and add each vitamin sequentially, allowing each to dissolve before the next: 10.2 mg biotin, 10.3 mg folic acid, 82 mg DL-pantothenic acid hemicalcium salt, 82 mg nicotinic acid, 82 mg riboflavin, 82 mg thiamine HCl, 82 mg pyridoxine HCl, 10.2 mg 4-aminobenzoic acid, and 10.3 mg cyanocobalamin. Bring to a final volume of 500 mL with Milli-Q H_2O and filter sterilize.
7. Reducing agent—Make directly before adding to the sterilized medium. Dissolve 1 g L-cysteine HCl and 1 g $Na_2S \bullet 9H_2O$ in 50 mL Milli-Q H_2O and filter sterilize.

3.2.2 Fermentation Medium

1. For 1 L of fermentation medium, add the following to 750 mL ddH_2O: 1.19 g trypticase, 0.71 g KCl, 0.71 g NaCl, 0.24 g $CaCl_2 \bullet 2H_2O$, 0.59 g $MgSO_4 \bullet 7H_2O$, 1.78 g PIPES, and 0.64 g NH_4Cl. To the basal medium, add 1.17 mL resazurin solution, 11.9 mL trace mineral solution, 11.9 mL hemin solution, and 11.9 mL fatty acid solution.
2. Adjust the pH of the medium to pH 6.8 using 3 M KOH. Bring the total volume to 1 L with Milli-Q H_2O and bubble with CO_2 for 3 hours to degas the medium.
3. Aliquot out 76 mL per serum bottle, degas the headspace for 20 min before capping and crimping the vials.
4. Prepare the sodium-carbonate + vitamin + phosphate mix by combining 60 mL of the sodium phosphate buffer with 15 mL of the vitamin + phosphate mix.
5. Using 1 mL and 5 mL syringes, add the fermentable carbon source of interest (arabinogalactan from larch wood or pectin from apple) to the serum vials at a final concentration of 1% w/v, 4 mL of the sodium carbonate + vitamin + phosphate mix, and 1 mL of the reducing agent mix. Equilibrate the serum vials in an anaerobic cabinet at 37 °C.

3.2.3 Inoculation, Fermentation, and Sampling

1. Use a Class 2 Microbial Safety cabinet to prepare a 10% fecal slurry using degassed phosphate-buffered saline (*see* **Note 10**).

2. Weigh out 5 g fecal material from a donor into a stomacher bag.
3. Dilute the material to a total of 50 g using degassed phosphate-buffered saline.
4. Place the bag into the Stomacher and homogenize at 230 rpm for 45 s.
5. Use a degassed Stripette to collect the liquid from between the inner and outer bag and transfer it into a degassed Falcon tube.
6. Use a syringe to inoculate the vials at a final concentration of 0.4% fecal slurry (3 mL) in the anaerobic cabinet.
7. Withdraw 1–5 mL for "time-zero" quantification of the microbial consortium and starting concentrations of metabolites.
8. Samples (1–5 mL) are collected from the vial using a sterile syringe at different timepoints every 4 to 8 hour intervals over 48–72 hours (*see* **Note 11**).
9. Sample aliquots (2 mL) from the colonic in vitro fermentation are pelleted by centrifugation (2800× *g* for 10 min at 4 °C) and frozen (−80 °C).
10. The supernatant is stored at −80 °C until metabolite analysis, and pellets are placed in lysing buffer for DNA isolation, like fecal samples as described in Chapter 5 (*see* **Note 12**).

3.3 Targeted Quantification of SCFAs and BCFAs by LC-MS/MS

3.3.1 Sample Preparation

1. Prepare the isotopically labeled fatty acid standards to 100 mM in MilliQ water (D_4-acetic acid, $^{13}C_2$-butyric acid, D_6-isobutyric acid, D_9-isovaleric acid, D_2-propionic acid, D_9-valeric acid).
2. Dilute the standards in 0.5% phosphoric acid-acidified MilliQ water to make a master mix containing 5 mM D_4-acetic acid, and 0.5 mM of each other standard. If isotopically labeled standards are unavailable (*see* **Note 13**).
3. Separately prepare quality control samples with all unlabeled fatty acids at 0.05 mM, 1 mM, and 8 mM final concentrations.
4. Mix 1:10 with the master mix containing the internal standards (10 μL of control sample to 90 μL of internal standard).
5. Thaw the supernatants from the in vitro colon model before centrifugation at 17000 g for 10 min to remove particulates, dilute the supernatant 1:10 with MilliQ water, and add diluted supernatant to internal standards. Routinely, 10 μL of 1:10 diluted supernatant is diluted with 90 μL of MilliQ water acidified with 0.5% phosphoric acid containing the fatty acid internal standards.
6. Aliquot samples and standards into HPLC vials with fixed inserts and place in the autosampler.

3.3.2 LC-MS/MS Analysis

1. Use the following settings for the Agilent 1290 HPLC system with autosampler fitted with a Thermo Scientific Hypercarb column with guard column:
2. Mobile phase A = 0.1% formic acid in water; Mobile phase B = 0.1% formic acid in acetonitrile.
3. Flow rate = 0.15 mL/min.
4. 2 μL injections of samples and standards
5. Gradient from 0% B to 60% B over 4 minutes, followed by 100% B for 2 min, and then re-equilibrate for the next injection with 0% B for 4 min.
6. Quantify the target fatty acids and their corresponding internal standard using the ion counts at the following multiple reaction monitoring (MRM) transitions and their respective retention times as shown in Table 13.2 (*see* **Note 14**).
7. After verifying the separation of the internal standards, graph the response ratio of each unlabeled standard with its respective isotopically labeled fatty acid and graph the standard curve for each metabolite for 0.05 mM, 1 mM, and 8 mM.
8. Calculate the original concentration of the fatty acid in the unknown samples by using the ratio of ion counts of the unknown to the internal standard (*see* **Note 15**).

Table 13.2
MRM transitions and retention times for SCFAs and BCFAs and their respective internal standard

Analyte	MRM transition	Retention time
Acetate	61.1 > 43.0	~1.5 min
Butyrate	89.1 > 43.1	~3.7 min
Isobutyrate	89.1 > 43.1	~2.9 min
Isovalerate	103.1 > 43.0	~4.2 min
Propionate	75.0 > 29.0	~2.3 min
Valerate	103.1 > 75.0	~4.7 min
D_4-acetate	65.1 > 47.0	~1.5 min
$^{13}C_2$-butyrate	91.1 > 44.0	~3.7 min
D_6-isobutyrate	95.0 > 49.0	~2.9 min
D_9-isovalerate	112.2 > 50.2	~4.2 min
D_2-propionate	77.0 > 31.1	~2.3 min
D_9-valerate	112.1 > 80.0	~4.7 min

9. Back-calculate the initial starting concentration of the sample if a dilution has been made to accurately quantify the fatty acids in the material.

3.4 NMR Metabolomics

3.4.1 NMR Buffer with Internal Standard

1. Dissolve 0.26 g NaH_2PO_4, 1.41 g K_2HPO_4, 100 mg NaN_3, and 17 mg TSP in 80 mL D_2O.
2. Adjust the pH to 7.4 with NaOH and/or HCl.
3. Bring the final volume to 100 mL with D_2O and store at 4 °C.

3.4.2 Urine and Colon Model Sample Preparation

1. Centrifuge urine samples at 17000 g for 10 min to remove any remaining particulates from the supernatant.
2. Thaw the in vitro colon model samples. Centrifuge samples at 17000 g for 10 min to remove particulates.
3. Mix 400 μL of the supernatant with 200 μL of the phosphate buffer with internal standard. Aliquot 500 μL of the mixture into a 5 mm NMR tube and keep the tubes at 4 °C until analysis.
4. Prepare a blank with 400 μL MilliQ water and 200 μL of the phosphate buffer with internal standard.

3.4.3 Fecal Extract Preparation

1. Thaw frozen fecal samples prepared using methods in Chapter 3 and resuspend at 20 mg feces/mL in phosphate buffer with internal standard.
2. Centrifuge samples at 17000 g for 10 min to remove particulates.
3. Aliquot 500 μL of the mixture into a 5 mm NMR tube and keep the tubes at 4 °C until analysis.
4. Prepare a blank with 500 μL of the phosphate buffer with internal standard.

3.4.4 NMR Setup and Acquisition

1. For adequate resolution, ^1H spectra are acquired using a 600 MHz NMR spectrometer set at 298 K with a pre-saturation approach such as 1D-NOESY (noesygppr1d) using 128 scans, a 12 ppm spectral width, a recycle delay of 2 to 4 s (D1), and a mixing time of 0.01 s (D8). The resulting acquisition time is about 2.6 s. The 90° pulse length is about 8.8 μs.
2. Spectra are transformed with 0.3 Hz line broadening and zero filling, manually phased, and baseline corrected using the TOP-SPIN software and the TSP signal set to 0 ppm.

3.4.5 NMR Data Analysis

Literature searches and in-house data acquisition have created libraries of over 80 compounds for urine (*see* **Note 16**) and over 45 compounds for fecal extract and colon model samples (*see*

Note 17). Two-dimensional NMR techniques can help further increase the number of identifiable compounds (*see* **Note 18**).

1. Using the Chenomx NMR Suite, manually integrate peaks for metabolites of interest by adjusting the peak height corresponding to the highest specific and unobstructed signal to match the raw spectrum. See Table 13.3 for relative chemical shifts for common SCFA, and others can be found in the Human Metabolome Database—https://www.hmdb.ca.
2. Repeat for all metabolites of interest amongst all samples to quantify their abundance.

Table 13.3
NMR multiplicity and relative chemical shifts for common SCFAs

Analyte	Multiplicity	Chemical shift
Acetate	Singlet	~2.0 ppm
Butyrate	Triplet Multiplet Triplet/multiplet	~0.9 ppm[a] ~1.55 ppm ~2.15 ppm
Isobutyrate	Doublet Multiplet	~1.05 ppm[a] ~2.3 ppm
Isovalerate	Doublet Multiplet Doublet	~0.9 ppm[a] ~1.9 ppm ~2.05 ppm
Propionate	Triplet Quartet	~1.05 ppm[a] ~2.17 ppm
Lactate	Doublet Quartet	~1.3 ppm ~4.1 ppm
Succinate	Singlet	~2.4 ppm
Valerate	Triplet Multiplet Multiplet Triplet	~0.9 ppm[a] 1.3 ppm ~1.5 ppm ~2.2 ppm
2-methyl-butyrate	Triplet Triplet Multiplet Multiplet Multiplet	~0.9 ppm[a] ~1.0 ppm ~1.4 ppm ~1.5 ppm ~2.2 ppm

[a]Potential for overlap in peaks

3.5 Multivariate Analysis of Targeted LC-MS and NMR Metabolomics Data

3.5.1 Initial Data Checks

1. Review data and remove samples if technical issues (e.g., faulty instrument, sample preparation error) have occurred on a particular sample or set of samples.
   - If technical issues occurred with a measurement platform, remove the data from the corresponding metabolites, unless the issue can be corrected (*see* **Note 19**).
2. Calculate the percentage of missing data for each metabolite (*see* **Notes 20-22**).
3. Conduct per-sample data scaling when gross differences in concentration between samples need to be corrected. This is accomplished for each sample by dividing all measured concentrations by a factor, for instance, the sum or the median of all concentrations for this sample.
4. Carry out an exploratory data analysis, such as principal component analysis (PCA), on the table of concentrations comprising all samples and metabolites from either NMR or LC/MS platform (*see* **Note 23**). Inspect the scores and loading plots, and the variance represented by each PC (*see* **Note 24**).
5. Investigate outliers, and if this highlights further issues with either samples or measurements that were not previously apparent, revisit previous steps. PCA is also useful as an unsupervised method of clustering samples (see below).

3.5.2 Analysis of the Pre-cleaned Metabolite Concentration Table

The data analysis strategy must be tailored to the research question (s) and design of the experiment. We refer the reader to Chapters 2 and 12 on study design and statistical analysis, respectively, for an in-depth discussion of these considerations and guidance on univariate analyses (*see* **Note 25**).

3.5.3 Determining Between-Group Differences

As the group membership is known, use a supervised method of classification (i.e., one in which the group membership is used to define the classification rule) such as PCA-DA (principal component analysis- discriminant analysis) or PLS-DA (partial least squares-discriminant analysis) (*see* **Note 26**), on the entire concentration table (*see* **Note 27**) as follows:

1. Split the observations into training and test sets (with, e.g., 2/3 data in the training set and 1/3 in the test set) (*see* **Note 28**).
2. Build the model with the training set (e.g., PCA-DA) and collate the model parameters (e.g., PCA loadings and group centroids and boundaries).
3. Apply the model parameters to the test set.
4. Infer the sensitivity and specificity of the method from the test set classification results (not those of the training set).

5. If interpretation is important, inspect the loadings of the axes on which discrimination is observed. Interpretation can be complemented by using other approaches (see below).

3.5.4 Identifying Between-Group Metabolite Differences

Univariate tests are carried out separately for each metabolite of interest. The results can then be examined in the context of underlying biological metabolic networks by carrying out pathway analysis. This requires the use of specialized software packages, such as MetaboAnalyst. The key steps for successful implementation are listed below:

1. Map the metabolite names to names/codes recognized by the software package (match names/ codes to its libraries) (*see* **Note 29**).
2. Collate the univariate analyses results for all metabolites (e.g., based on t-test or equivalent nonparametric test) and use them as input in the pathway analysis.
3. Select a relevant pathway library (for instance, homo sapiens SMPDB) (*see* **Note 29**).

3.5.5 Relating a Quantitative Measure to a Panel of Metabolite Concentrations

The relationship between a quantitative variable and a panel of metabolites can be investigated using multiple regression analysis. However, multi-collinearity between metabolites (explanatory variables) can lead to issues when using multiple regression:

1. Use PCR (principal component regression) or PLS (partial least squares) regression to model the quantitative measure as a function of latent variables which are based on the metabolite concentrations, rather than the concentrations themselves (as is the case in multiple regression). This avoids multi-collinearity issues, making it possible to include all metabolites as explanatory variables.
2. Use RMSEP (root-mean-square error of prediction) to evaluate regression fit (*see* **Note 30**) and inspect loadings for interpretation (*see* **Note 31**).

3.5.6 Analyzing Time Courses

With longitudinal and time-course studies, the samples are not independent. The between-individual (or colon model) variability can mask the effect of the factors that are under study. The data analysis strategies described above can be adapted to take this into account.

Use multi-level versions of PCA, PCA-DA, or PLS-DA, offered in the R mixOmics software package, to integrate into the analysis a step which separates the individual (or per-colon model time course) variance from that of other factors.

4 Notes

1. Urine sample collection is noninvasive and easy making it an important sample for dietary intervention studies. The primary constituent of urine is water, and the remaining ~5% is urea, organic and inorganic salts, and small molecule metabolites (less than 2000 Da). Over 3000 metabolites have been reported to be identifiable in human urine (Urine Metabolome database—https://urinemetabolome.ca) [10]. In practice, up to a few hundred may be detected with a fully implemented analytical framework. Until recently, urine was understood to be sterile but with improvements to nucleic acid isolation techniques and sequencing technologies, samples with low biomass or low abundance in bacteria can be readily sequenced [11–13].
2. Colon models are a simple and reliable tool to estimate the contribution of the gut microbiome to the catabolism of dietary substrates. Using a semi-defined PIPES-buffered medium under anaerobic conditions with ammonia as a nitrogen source can allow for targeted analysis of the contribution of culturable microorganisms in the fermentation of different carbon sources [14–19]. Additionally, providing starting amounts of SCFAs, BCFAs, hemin, and a mixture of B-vitamins allows for the cultivation of more fastidious commensal microorganisms, which normally acquire these nutrients from the diet, host, or other members of the gut microbiome [20–22]. The model can also be readily modified to investigate how the microbiome contributes to other metabolic processes (e.g., acquiring nitrogen or exchanging vitamins within the microbial community) by eliminating the nutrient(s) of interest and monitoring the shift in the composition of the community and the production of fermentation products.
3. Targeted metabolomics quantifies metabolites in biological samples to generate more complete models of metabolism in a variety of cells and microbial communities [23, 24]. LC-MS and NMR can be used in parallel to gain a better understanding of biological systems. Quantification of metabolites by LC-MS is favored because of the sensitivity of the instrumentation, which can reliably quantify metabolites in the pM to nM range depending on the sample. However, issues such as ion suppression and the matrix used make absolute quantification between samples complex, as it is both sample- and matrix-dependent. Using isotopically labeled internal standards (^{2}H, ^{15}N, or ^{13}C-labeled metabolites of interest) allows for differences in ionization to be corrected. At scale, this can be applied to accurately quantify in the order of 200–400 metabolites per sample with the use of internal standards [8]. However, the

required upfront method development and standardization add cost and complexity for routine analysis. NMR benefits from the signal intensity obtained being proportional to the concentration of the metabolite of interest. NMR can reliably quantify metabolites in solution in the μM to mM range. Through literature searches and method development, quantification of 50–100 metabolites per sample is achievable, making this a robust method for identifying gross changes between samples.

4. Although quartz NMR tubes offer the best quality and precision, standard borosilicate NMR will provide sufficient resolution and is inexpensive way to support the relatively high-throughput nature of our methodology. NMR tubes can be reused by washing in deionized water, soaking in acetone for 1 hour, and then triple rinsing with deionized water. NMR tubes can be washed with a vacuum-assisted NMR tube-cleaning apparatus.
5. A 24 h urine collection involves the collection of all voided urine over a 24-hour period. The very first morning void is discarded but marks the start of the 24-hour urine collection. The second void is collected and then continuously until the next day where the first void is included and marks the end of the 24-hour urine collection.
6. Metabolites in 24 hour urine collection without preservatives when kept ≤4 °C for 24 hours or less are stable [25]. However, if samples need to be stored at ambient temperature or for longer than 48 hours, thymol is recommended. Thymol (5 mL, 10% thymol w/v in isopropanol/liter of urine) effectively inhibits bacterial growth and does not cause any shifts in metabolic profiles in urine analyzed by MS [26]. Boric acid at 200 mM was reported to be bacteriostatic and to not affect NMR metabolic signatures but may impact electrospray ionization in metabolomic analysis using MS [27]. Urine pH differences cause shifts in NMR spectra [28].
7. There is significant inter- and intraindividual variability in urine volume and density of urine particulates. Several methods can normalize urine to minimize variability in urine samples. The most reliable is osmolality-based normalization [29, 30], but access to an osmometer may be limited or time-consuming if the sample needs to be sent to a clinical lab for analysis. However, a more cost-effective method of normalization is normalizing using urine specific gravity or simple total volume excreted over a 24-hour period [31]. Normalizing urine using urinary creatinine has been regarded as the gold standard, but creatine concentrations vary due to participant age and gender, hence can introduce significant bias into the metabolomic quantification [32].

8. Ensure that an appropriate volume of urine is aliquoted into tubes to minimize dead volume and leakage of the sample due to expansion of urine volume during freezing.
9. Fecal inoculum (slurry) should be prepared with fresh (less than 2 hours after excretion) or frozen samples, which have been stored with cryopreservation (10–20% glycerol) [16]. If colon model facilities are not available, metabolites can be measured directly in the fecal slurry.
10. All plastics used in the inoculation and processing of samples during the in vitro fermentation should be equilibrated in an anaerobic cabinet for 1 week to eliminate the oxygen present in pipette tips, syringes, stomacher bags, conical tubes, and microcentrifuge tubes. This is important to reduce the effect of introducing environmental oxygen during the fermentation and retain the viability of oxygen-sensitive microorganisms. Degas the phosphate-buffered saline by equilibrating it in the anaerobic cabinet for 1 week before use and minimize oxygen exposure by tightening the lid before removing from the anaerobic workstation.
11. The greatest increase in metabolites (SCFA and BCFA) occurs during the first 12–24 hours; therefore, the study design for sampling should be carefully planned [16].
12. Purified genomic DNA can be used for 16S or metagenomic sequencing to monitor shifts in the microbial community over time (see Chapters 6 and 7).
13. We recommend using isotopically labeled internal standards for quantifying metabolites by LC-MS/MS. However, if unavailable, a matrix-matched calibration curve with unlabeled standards can be used. For an in vitro colon model, a matrix-matched calibration curve can be produced by adding 0.001 mM to 10 mM of unlabeled standard to a blank of uninoculated fermentation medium acidified with 0.5% phosphoric acid. These standards should be treated exactly like biological samples by following the sample preparation process and quantifying the signal corresponding to the metabolite of interest. The concentration of the SCFAs in the unknown samples can be interpolated from the ion counts from this standard curve within its linear range.
14. Although the MRM transitions are metabolite specific, variations in retention time may occur depending on HPLC, column, flow rate, and gradient used. Verify retention times using standards.
15. The limit of detection of most fatty acids is 1 μM [33]. The limit of quantification for most fatty acids is 3 μM and 9 μM for acetate.

16. The human urinary ^{1}H NMR spectrum is dominated by the signals of creatine; creatinine; urea; glycine, citric, acetic, and hippuric acids; alanine; and trimethylamine-N-oxide [10]. Many amino acids, some amines, tricarboxylic acid cycle, and glycolysis compounds are also detectable, notably, serine, taurine, or the branched amino acids, leucine, isoleucine and valine and their products, 2-hydroxyisovalerate, 2-hydroxyvalerate, methylsuccinate, or 2-oxoisocaproate. Succinate and fumarate and pyruvate, lactate, or formate are also readily detected. Sugars and sugar alcohols are found in small amounts (glucose, arabinose, fucose, myo-inositol, mannitol to name a few) and, interestingly, the signals of microbial products such as dimethylamine, indoxyl sulfate, hippurate, and phenylacetylglutamine are easily seen due to their position in less-crowded parts of the 1D spectrum, making their quantification relatively straightforward. Interestingly, butyrate, isobutyrate, and isovalerate have been detected in the NMR urine spectra of volunteers; however, in practice, only isobutyrate is regularly reported in publications (Le Gall, unpublished) [34, 35].
17. Compounds detectable in fecal extract and colon model samples include most amino acids, SCFAs, betaine, trehalose, and fermentation products including ethanol, formate, and acetaldehyde [36, 37]
18. To aid in the identification of metabolites, two-dimensional NMR techniques such as COSY, HSQC, and HMBC can be employed to further characterize compounds in samples that exhibit signals of interest in their 1D spectra. The analysis can then proceed with the analysis of the 2D NMR spectra complemented with a thorough literature search, the use of the Urine Metabolome and the Human Metabolome Databases (https://urinemetabolome.ca; https://www.hmdb.ca) and Chenomx NMR Suite to build a relevant in-house metabolite list.
19. This step may be repeated further in the workflow if there anomalies (typically, missing values or gross outliers) have occurred.
20. For metabolites whose absence is above a certain threshold (suggested 20%), discard the metabolite(s) from the general workflow and carry out a separate analysis to determine if the missing values are related to the biological factor(s) under study (*see* **Note 21**). In metabolomics data, missing values are normally due to metabolite concentrations being below the limit of detection. The following recommendations derive from this assumption. For those whose percent missing is below the threshold, impute missing values (*see* **Note 22**) and retain them in the general workflow. For the imputation, check

that data missings is due to concentrations being below the detection limit. This being the case, impute missing data for each metabolite separately, to a fraction of the minimal value that is detected for this metabolite (a fixed fraction such as 0.5× the minimal value, or a fraction randomly chosen for each sample). If otherl reasons exist that led to data missing, a different approach may have to be followed, being tailored to the situation.

21. For metabolites removed from the workflow due to a high proportion of missing values, test whether there is a relationship between the absence of values and an experimental factor. When the biological factor of interest is categorical (e.g., different treatments or experimental conditions), calculate the percentage missing for each group and use a Fisher's exact test to compare the proportion of missing values between groups. Use a regression for quantitative factors.
22. Some commercially acquired data automatically impute missing data. Check the documentation provided to understand how the data was derived.
23. PCA is offered in numerous software packages, though not all are able to handle situations in which there are more measurements (metabolites) than observations (samples).
24. In metabolomics datasets, it is not uncommon for the distribution of a metabolite's concentrations to have a long right tail (a small number of values which are much larger than their counterparts), with the rest being normally distributed. Check for atypically high PCA scores and loadings, which are often indicative of a limited number of samples and metabolites exhibiting this behavior.
25. Data issued from the two analytical platforms should be analyzed separately. This is not an absolute requisite but can facilitate interpretation and lessen the chance of encountering issues linked to, e.g., measurement error distribution. We focus on approaches for data analysis that specifically consider the multivariate nature of the data and acknowledge that metabolites are part of a complex biological network.
26. PLS-DA and PCA-DA both carry out a data reduction step, on which DA is then based. The fundamental difference between them is that PLS-DA uses the group membership information for the data reduction step, whereas PCA-DA only uses the metabolomics table. Consequently, discrimination may appear better for the training set with PLS-DA. However, only the results from a correctly validated set reflect the true ability to distinguish the groups of interest.

27. If the observations are not independent, include all samples from a set of non-independent observations either in the same training or the same validation segment.
28. For small datasets, use a leave-one-out or leave-a-segment-out validation, a procedure in which one observation (or a segment comprising a few observations) is removed to serve as a test set, with the rest being used for training. The procedure is repeated for each observation (or segment) in turn, and the classification performance of all test sets is aggregated to compute the validation classification results.
29. For pathway analysis, mapping is based on HMDB IDs, KEGG IDs, or metabolite names. Use the format(s) that maximize the number of metabolites that are mapped, as unmapped metabolites are discarded from the analysis. For this, test whether a combination, or only one type of metabolite names/codes, can be used. Use the MetaboAnalyst compound ID conversion tool to obtain compound names and/or codes listed in the software libraries.
30. Use a validation procedure (typically, internal cross-validation) to evaluate the prediction results.
31. When using many metabolites as explanatory variables, interpretation of the models can be challenging. Use sparse versions of PLS-regression if interpretation, rather than prediction ability, is the key objective (offered in the R package mixOmics).

Acknowledgements

This work was funded by Biotechnology and Biological Sciences Research Council (BBSRC) through Institute Strategic Programme awards Gut Microbes and Health BB/R012490/1 and Food, Microbiome and Health [BB/X011054/1] and their constituent project(s), and by BBSRC Core Capability Grants BB/CCG1860/1 and BB/CCG2260/1.

References

1. Human Microbiome Project C (2012) Structure, function and diversity of the healthy human microbiome. Nature 486(7402): 207–214. https://doi.org/10.1038/nature11234
2. Visconti A, Le Roy CI, Rosa F, Rossi N, Martin TC, Mohney RP, Li W, de Rinaldis E, Bell JT, Venter JC, Nelson KE, Spector TD, Falchi M (2019) Interplay between the human gut microbiome and host metabolism. Nat Commun 10(1):4505. https://doi.org/10.1038/s41467-019-12476-z
3. Koh A, De Vadder F, Kovatcheva-Datchary P, Backhed F (2016) From dietary fiber to host physiology: short-chain fatty acids as key bacterial metabolites. Cell 165(6):1332–1345. https://doi.org/10.1016/j.cell.2016.05.041
4. Martin AM, Sun EW, Rogers GB, Keating DJ (2019) The influence of the gut microbiome on host metabolism through the regulation of

gut hormone release. Front Physiol 10:428. https://doi.org/10.3389/fphys.2019.00428

5. Yano JM, Yu K, Donaldson GP, Shastri GG, Ann P, Ma L, Nagler CR, Ismagilov RF, Mazmanian SK, Hsiao EY (2015) Indigenous bacteria from the gut microbiota regulate host serotonin biosynthesis. Cell 161(2):264–276. https://doi.org/10.1016/j.cell.2015.02.047
6. Zeng H, Umar S, Rust B, Lazarova D, Bordonaro M (2019) Secondary bile acids and short chain fatty acids in the colon: a focus on colonic microbiome, cell proliferation, inflammation, and cancer. Int J Mol Sci 20(5). https://doi.org/10.3390/ijms20051214
7. Portincasa P, Bonfrate L, Vacca M, De Angelis M, Farella I, Lanza E, Khalil M, Wang DQ, Sperandio M, Di Ciaula A (2022) Gut microbiota and short chain fatty acids: implications in glucose homeostasis. Int J Mol Sci 23(3). https://doi.org/10.3390/ijms23031105
8. Nemkov T, Hansen KC, D'Alessandro A (2017) A three-minute method for high-throughput quantitative metabolomics and quantitative tracing experiments of central carbon and nitrogen pathways. Rapid Commun Mass Spectrom 31(8):663–673. https://doi.org/10.1002/rcm.7834
9. Emwas A-HM (2015) The strengths and weaknesses of NMR spectroscopy and mass spectrometry with particular focus on metabolomics research. In: Bjerrum JT (ed) Metabonomics: methods and protocols. Springer, New York, pp 161–193. https://doi.org/10.1007/978-1-4939-2377-9_13
10. Bouatra S, Aziat F, Mandal R, Guo AC, Wilson MR, Knox C, Bjorndahl TC, Krishnamurthy R, Saleem F, Liu P, Dame ZT, Poelzer J, Huynh J, Yallou FS, Psychogios N, Dong E, Bogumil R, Roehring C, Wishart DS (2013) The human urine metabolome. PLoS One 8(9):e73076. https://doi.org/10.1371/journal.pone.0073076
11. Hilt EE, McKinley K, Pearce MM, Rosenfeld AB, Zilliox MJ, Mueller ER, Brubaker L, Gai X, Wolfe AJ, Schreckenberger PC (2014) Urine is not sterile: use of enhanced urine culture techniques to detect resident bacterial flora in the adult female bladder. J Clin Microbiol 52(3):871–876. https://doi.org/10.1128/JCM.02876-13
12. Pearce MM, Hilt EE, Rosenfeld AB, Zilliox MJ, Thomas-White K, Fok C, Kliethermes S, Schreckenberger PC, Brubaker L, Gai X, Wolfe AJ (2014) The female urinary microbiome: a comparison of women with and without urgency urinary incontinence. MBio 5(4): e01283-01214. https://doi.org/10.1128/mBio.01283-14
13. Pohl HG, Groah SL, Perez-Losada M, Ljungberg I, Sprague BM, Chandal N, Caldovic L, Hsieh M (2020) The urine microbiome of healthy men and women differs by urine collection method. Int Neurourol J 24(1):41–51. https://doi.org/10.5213/inj.1938244.122
14. Williams BA, Bosch MW, Boer H, Verstegen MWA, Tamminga S (2005) An in vitro batch culture method to assess potential fermentability of feed ingredients for monogastric diets. Anim Feed Sci Technol 123-124:445–462. https://doi.org/10.1016/j.anifeedsci.2005.04.031
15. Jonathan MC, van den Borne JJGC, van Wiechen P, Souza da Silva C, Schols HA, Gruppen H (2012) In vitro fermentation of 12 dietary fibres by faecal inoculum from pigs and humans. Food Chem 133(3):889–897. https://doi.org/10.1016/j.foodchem.2012.01.110
16. Colosimo R, Harris HC, Ahn-Jarvis J, Troncoso-Rey P, Finnigan TJA, Wilde PJ, Warren FJ (2024) Colonic in vitro fermentation of mycoprotein promotes shifts in gut microbiota, with enrichment of Bacteroides species. Commun Biol 7(1):272. https://doi.org/10.1038/s42003-024-05893-4
17. Ravi A, Troncoso-Rey P, Ahn-Jarvis J, Corbin KR, Harris S, Harris H, Aydin A, Kay GL, Le Viet T, Gilroy R, Pallen MJ, Page AJ, O'Grady J, Warren FJ (2022) Hybrid metagenome assemblies link carbohydrate structure with function in the human gut microbiome. Commun Biol 5(1):932. https://doi.org/10.1038/s42003-022-03865-0
18. Koev TT, Harris HC, Kiamehr S, Khimyak YZ, Warren FJ (2022) Starch hydrogels as targeted colonic drug delivery vehicles. Carbohydr Polym 289:119413. https://doi.org/10.1016/j.carbpol.2022.119413
19. do Prado SBR, Minguzzi BT, Hoffmann C, Fabi JP (2021) Modulation of human gut microbiota by dietary fibers from unripe and ripe papayas: distinct polysaccharide degradation using a colonic in vitro fermentation model. Food Chem 348:129071. https://doi.org/10.1016/j.foodchem.2021.129071
20. Rodionov DA, Arzamasov AA, Khoroshkin MS, Iablokov SN, Leyn SA, Peterson SN, Novichkov PS, Osterman AL (2019) Micronutrient requirements and sharing capabilities of the human gut microbiome. Front Microbiol 10:1316. https://doi.org/10.3389/fmicb.2019.01316

21. Sharma V, Rodionov DA, Leyn SA, Tran D, Iablokov SN, Ding H, Peterson DA, Osterman AL, Peterson SN (2019) B-vitamin sharing promotes stability of gut microbial communities. Front Microbiol 10:1485. https://doi.org/10.3389/fmicb.2019.01485
22. Culp EJ, Goodman AL (2023) Cross-feeding in the gut microbiome: ecology and mechanisms. Cell Host Microbe 31(4):485–499. https://doi.org/10.1016/j.chom.2023.03.016
23. Bennett BD, Kimball EH, Gao M, Osterhout R, Van Dien SJ, Rabinowitz JD (2009) Absolute metabolite concentrations and implied enzyme active site occupancy in Escherichia coli. Nat Chem Biol 5(8): 593–599. https://doi.org/10.1038/nchembio.186
24. Park JO, Rubin SA, Xu YF, Amador-Noguez D, Fan J, Shlomi T, Rabinowitz JD (2016) Metabolite concentrations, fluxes and free energies imply efficient enzyme usage. Nat Chem Biol 12(7):482–489. https://doi.org/10.1038/nchembio.2077
25. Rotter M, Brandmaier S, Prehn C, Adam J, Rabstein S, Gawrych K, Bruning T, Illig T, Lickert H, Adamski J, Wang-Sattler R (2017) Stability of targeted metabolite profiles of urine samples under different storage conditions. Metabolomics 13(1):4. https://doi.org/10.1007/s11306-016-1137-z
26. Wang X, Gu H, Palma-Duran SA, Fierro A, Jasbi P, Shi X, Bresette W, Tasevska N (2019) Influence of storage conditions and preservatives on metabolite fingerprints in urine. Meta 9(10). https://doi.org/10.3390/metabo9100203
27. Roux A, Thevenot EA, Seguin F, Olivier MF, Junot C (2015) Impact of collection conditions on the metabolite content of human urine samples as analyzed by liquid chromatography coupled to mass spectrometry and nuclear magnetic resonance spectroscopy. Metabolomics 11(5):1095–1105. https://doi.org/10.1007/s11306-014-0764-5
28. Huart J, Leenders J, Taminiau B, Descy J, Saint-Remy A, Daube G, Krzesinski JM, Melin P, de Tullio P, Jouret F (2019) Gut microbiota and fecal levels of short-chain fatty acids differ upon 24-hour blood pressure levels in men. Hypertension 74(4):1005–1013. https://doi.org/10.1161/HYPERTENSIONAHA.118.12588
29. Gagnebin Y, Tonoli D, Lescuyer P, Ponte B, de Seigneux S, Martin PY, Schappler J, Boccard J, Rudaz S (2017) Metabolomic analysis of urine samples by UHPLC-QTOF-MS: impact of normalization strategies. Anal Chim Acta 955: 27–35. https://doi.org/10.1016/j.aca.2016.12.029
30. Chetwynd AJ, Abdul-Sada A, Holt SG, Hill EM (2016) Use of a pre-analysis osmolality normalisation method to correct for variable urine concentrations and for improved metabolomic analyses. J Chromatogr A 1431:103–110. https://doi.org/10.1016/j.chroma.2015.12.056
31. Edmands WM, Ferrari P, Scalbert A (2014) Normalization to specific gravity prior to analysis improves information recovery from high resolution mass spectrometry metabolomic profiles of human urine. Anal Chem 86(21): 10925–10931. https://doi.org/10.1021/ac503190m
32. Rosen Vollmar AK, Rattray NJW, Cai Y, Santos-Neto AJ, Deziel NC, Jukic AMZ, Johnson CH (2019) Normalizing untargeted periconceptional urinary metabolomics data: a comparison of approaches. Meta 9(10). https://doi.org/10.3390/metabo9100198
33. Saha S, Day-Walsh P, Shehata E, Kroon PA (2021) Development and validation of a LC-MS/MS technique for the analysis of short chain fatty acids in tissues and biological fluids without derivatisation using isotope labelled internal standards. Molecules 26(21). https://doi.org/10.3390/molecules26216444
34. Schmedes M, Aadland EK, Sundekilde UK, Jacques H, Lavigne C, Graff IE, Eng Ø, Holthe A, Mellgren G, Young JF, Bertram HC, Liaset B, Clausen MR (2016) Lean-seafood intake decreases urinary markers of mitochondrial lipid and energy metabolism in healthy subjects: metabolomics results from a randomized crossover intervention study. Mol Nutr Food Res 60(7):1661–1672. https://doi.org/10.1002/mnfr.201500785
35. Carrola J, Rocha CM, Barros AS, Gil AM, Goodfellow BJ, Carreira IM, Bernardo J, Gomes A, Sousa V, Carvalho L, Duarte IF (2011) Metabolic signatures of lung cancer in biofluids: NMR-based metabonomics of urine. J Proteome Res 10(1):221–230. https://doi.org/10.1021/pr100899x
36. Kellingray L, Le Gall G, Doleman JF, Narbad A, Mithen RF (2021) Effects of in vitro metabolism of a broccoli leachate, glucosinolates and S-methylcysteine sulphoxide on the human faecal microbiome. Eur J Nutr 60(4):2141–2154. https://doi.org/10.1007/s00394-020-02405-y

37. Le Gall G, Noor SO, Ridgway K, Scovell L, Jamieson C, Johnson IT, Colquhoun IJ, Kemsley EK, Narbad A (2011) Metabolomics of fecal extracts detects altered metabolic activity of gut microbiota in ulcerative colitis and irritable bowel syndrome. J Proteome Res 10(9): 4208–4218. https://doi.org/10.1021/pr2003598

Chapter 14

Virome Analysis

Revathy Krishnamurthi, Rik Haagmans, Ryan Cook, Alise J. Ponsero, and Evelien M. Adriaenssens

Abstract

Viruses are omnipresent and the most abundant biological entities on the planet. They play a critical role in shaping the ecology and evolution of all life forms. This chapter provides methods to perform viromics or viral metagenomics, starting with the creation of a mock community to control experimental variables, followed by methods for nucleic acid extraction, and finally bioinformatics workflows for the processing and analysis of the sequence data.

Key words Viromics, Viral metagenomics, Mock community, Viral enrichment, Viral mining tools, Taxonomic annotation, Functional annotation

1 Introduction

Viromics (also known as viral metagenomics) provides a culture-independent approach to studying entire viral communities across diverse ecosystems. By circumnavigating laboratory cultivation, viromics enables the discovery of large numbers of Uncultivated Viral Genomes (UViGs), which are now being placed within existing taxonomic frameworks [1, 2]. A key difference between metagenomics (see Chap. 8) and viromics is the separation of the cellular fraction from the viral fraction in viromics, with the latter exclusively focused on sequencing the viral fraction. Recent studies have shown that viromics and metagenomics can recover distinct populations of viruses, with viromics being superior at capturing extracellular viruses [3, 4].

To evaluate laboratory methods for VLP enrichment and sequencing, a combination of viruses can be used as a mock virus community (MC). Quantification of the virus stocks used to build an MC can be performed using epifluorescence microscopy (EFM), rather than plaque assays, as this includes noninfectious particles containing viral nucleic acid which will be present in the sequencing data. As increased sensitivity is required for viruses with short

Simon R. Carding (ed.), *Best Practice in Microbiome Research*, Springer Protocols Handbooks,
https://doi.org/10.1007/978-1-0716-5009-7_14,

Table 14.1
Mock community virus information

Virus name	Genome nucleic acid	Topology	Length (kb)	Virion enveloped	Size (nm)	Accession
Prokaryotic						
Det7	dsDNA	Linear	157.5	No	90^{c}, 110^{t}	NC_027119.1
M13	ssDNA	Circular	6.4	No	6.5^{d}, 860^{l}	NC_003287.2
P22	dsDNA	Linear	41.7	No	60^{c}	NC_002371.2
Qbeta	ssRNA	Linear	4.2	No	25^{c}	NC_001890.1
T5	dsDNA	Linear	121.7	No	90^{c}, 160^{t}	NC_005859.1
Eukaryotic						
BVDV-1	ssRNA	Linear	12.5	Yes	50	NC_001461.1
MHV-68	dsDNA	Linear	119.5	Yes	220	NC_001826.2
RV-A	dsRNA	Linear (11 sg.)	18.6	No	80	NC_011507.2, NC_011506.2, NC_011508.2, NC_011510.2, NC_011500.2, NC_011509.2, NC_011501.2, NC_011502.2, NC_011503.2, NC_011504.2, NC_011505.2
SINV	ssRNA	Linear	11.7	Yes	70	NC_001547.1

[c]capsid size, [t]tail length, [d]virion diameter, [l]virion length, *sg* genome segments

single-stranded genomes, compared to double-stranded DNA phages, the use of blank samples is important to ensure the absence of VLPs in the media. The protocol below has quantified various viruses as members of an MC, including the bacterial viruses Det7, M13, P22, T5, Qbeta, and eukaryotic viruses, including Bovine viral diarrhea virus 1, Murid gammaherpesvirus 4, Rotavirus A, and Sindbis virus (Table 14.1). The protocols, depicted in Fig. 14.1, cover three main considerations in analyzing fecal viromes: the development and quantification of an MC of viruses used as positive controls, total nucleic acid extraction of the viral fraction, and virome-specific bioinformatics processing of sequence data.

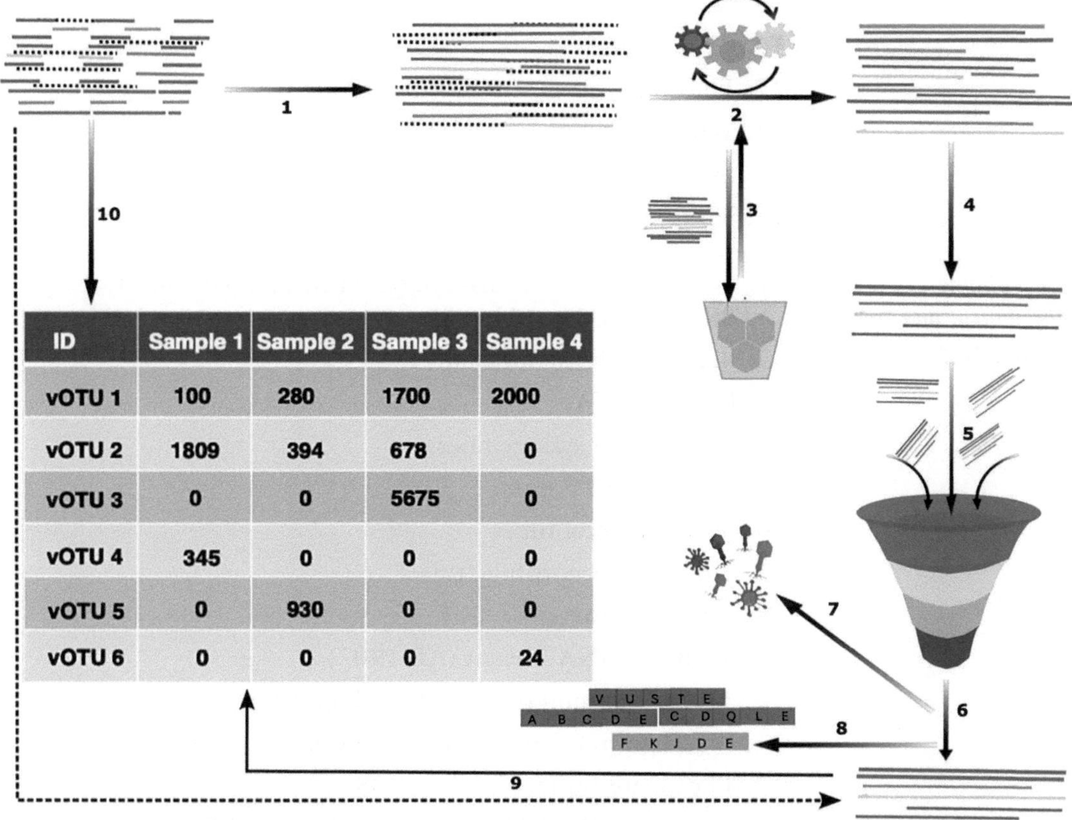

Fig. 14.1 Virome analysis pipeline describing (*1*) assembling raw reads using chosen assembly tools; (*2*) mining virome sequences from the assembled contigs using multiple viral mining tools; (*3*) optional binning; (*4*) dereplication of viral contigs to reduce the dataset complexity and improve read mapping; (*5*) filtering vOTUs for medium and above quality sequences with additional length-based thresholds; (*6*) final representative vOTUs; (*7*) taxonomic annotation; (*8*) functional annotation; (*9*) estimating the abundance of vOTUs in each sample under study by mapping the raw reads to the vOTUs; (*10*) assigning sample metadata and other genome-based annotation to create vOTU table for downstream data analysis

2 Materials

2.1 EFM Quantification of Mock Community Viruses

1. Class II Microbiological Safety Cabinet (MSCII).
2. Vacuum pump.
3. Vacuum trap (Buchner flask).
4. Swinnex filter holders.
5. Silicone gasket.
6. Whatman Anodisc 13-mm 0.02-μm pore filter membranes.
7. SYBR Gold.
8. Tweezers with fine tips.
9. Fluorescence microscope with a 100× objective.

10. Microscope slides.
11. Coverslips.
12. Fluoromount-G anti-fade mounting media.
13. Whatman filter paper number 4.
14. Sterile nuclease-free water (Invitrogen Catalog: 10977035).
15. Virus stock (Table 14.1).

2.2 Virome Enrichment and Total Nucleic Acid Extraction

1. 20× Homemade DNAse buffer (1 M Tris, 100 mM $CaCl_2$, and 30 mM $MGCl_2$, pH = 8)
2. 0.45 μm PES syringe filters
3. 0.5 M EDTA
4. 0.8 μm PES syringe filters
5. 10 mM dNTP mix
6. 50 mL conical tubes
7. 2 mL microcentrifuge tubes
8. Absolute ethanol.
9. TURBO DNA-free kit (AM1907).
10. Mock Community (*see* Subheading 2.1).
11. Nuclease-free water (Invitrogen Catalog: 10977035).
12. Phosphate-buffered saline (PBS).
13. Qubit™ dsDNA HS Assay Kit (Thermo Fisher Scientific Catalog: Q32851).
14. Sigma-Aldrich WTA2 Complete Whole Transcriptome Amplification Kit (WTA2-50RXN).
15. Qiagen QIAamp Viral RNA Mini Kit (52904).
16. Qiagen QIAquick PCR Purification Kit (28104).

2.3 Computational Tools for Viromics Data Processing and Analysis

1. BBTools [5] (*see* **Note 1**).
2. BEDTools [6].
3. Blast+ [7].
4. Bowtie2 v2.4.4 [8].
5. BWA-MEM [9].
6. CheckV and accessory scripts [10].
7. DeepVirFinder [11].
8. DIAMOND [12].
9. Fastp [13].
10. FastQC v0.11.9 (https://www.bioinformatics.babraham.ac.uk/projects/fastqc/).
11. geNomad [14].

12. ggplot2 v3.3.6 [15].
13. HH-suite [16].
14. iToL [17].
15. MEGAHIT [18].
16. MMseqs2 [19].
17. Pharokka v1.7.4 [20].
18. Python (https://www.python.org/).
19. R version 4.1.2 (https://www.r-project.org/).
20. SAMtools [21].
21. metaSPAdes v4.0.0 [22].
22. Trimmomatic v0.39 [23].
23. VIBRANT v1.2.1 [24].
24. ViptreeGen v1.1 [25].
25. VirSorter2 v2.2.4 [26].

3 Methods

3.1 Quantifying Virus Stocks Using Epifluorescence Microscopy

1. Prepare 800 μL 50× SYBR Gold by combining 4 μL 10,000× SYBR Gold with 800 μL H_2O.
2. Prepare 400 μL virus stock at 3×10^6 to 1×10^7 particles/ml in H_2O (*see* **Note 2**).
3. Combine 400 μL 50× SYBR Gold and 400 μL diluted virus stock and mix by pipetting.
4. To generate a blank sample, combine 400 μL 50× SYBR Gold and 400 μL H_2O and mix by pipetting (*see* **Note 3**).
5. Store the sample and the blank in the dark for 1 h.
6. Using tweezers, place the Anodisc filter onto the screen of the Swinnex filter holder outlet, making sure the filter is well-centered.
7. Screw on the filter holder inlet, carefully applying just enough torque to achieve a good seal without breaking the Anodisc membrane.
8. To ensure a good seal, apply vacuum pressure, add 300 μL water, and observe the flow rate. If the water is aspirated in <1–2 s, the membrane may have cracked.
9. If a good seal has been achieved, slowly add the sample to the inlet.
10. Wash the membrane once by passing through 300 μL of water.
11. Maintain vacuum for an additional 30 s once liquid has filtered through.

12. Disassemble the filter holder and then stop the vacuum pump.
13. Once pressure has equalized, remove the Anodisc membrane using tweezers.
14. Place the Anodisc membrane on a sheet of Whatman paper to dry in the dark for 1 min.
15. Using tweezers, keeping the membrane sample-side up, drop 15 μL Fluoromount-G onto a microscope slide.
16. Place the Anodisc membrane VLP-side up onto the mountant droplet using tweezers.
17. Add a 15 μL drop of Fluoromount-G onto the Anodisc membrane.
18. Gently place a coverslip over the membrane and gently compress.
19. Allow the mountant to set at 20 °C in the dark for 2 h.
20. Place the sample slide onto the microscope.
21. Focus the microscope using low magnification bright field illumination: place the edge of the disc into the center of the view of the objective with the smallest magnification, and make sure the edge of the membrane is in focus. Then, switch to objectives with increasing magnification, ensuring the edge of the membrane is in focus at each step, finishing with the 100× objective.
22. Switch to 488-nm fluorescent illumination and move the viewing field slightly towards the center of the membrane, until virus particles come into view. Make sure all particles in the view are in focus.
23. Determine the ideal exposure time and illumination strength for the sample by taking an image and inspecting the histogram, making sure that the peak of the histogram is around the middle of the brightness range of the image.
24. Take 15–20 images at random locations on the membrane using the same exposure settings for each image.
25. Remove the slide and clean up any remaining oil on the slide.
26. Place the blank and repeat the steps above from **steps 21** to **25**, then determine whether the blank contains any contaminating particles by taking ~10 images at the exposure settings used for the sample at random locations on the membrane.
27. Remove the blank and clean up any remaining oil from the slide.
28. Count the number of particles in each view of a filter (*see* **Note 4**).
29. Calculate the average number of particles.

30. Retrieve the dimensions of the view (*see* **Note 5**).
31. Calculate the sample particle titer:

$$\text{Views on filter} = \frac{\langle \text{Filter radius [mm]} \rangle^2 \times \pi}{\langle \text{View area [mm}^2] \rangle}$$

$$\text{Concentration [p/mL]} = \frac{\langle \text{Avg count} \rangle \times \langle \text{Views on filter} \rangle}{\langle \text{Sample volume [mL]} \rangle} \times \langle \text{Dilution factor} \rangle$$

For example, a virus stock has been previously estimated to contain 10^9 PFU/mL and is diluted 1:100 before staining. A volume of 0.02 mL of the 1:100 diluted stock is stained and filtered on a 13-mm filter (*see* **Note 6**):

$$\text{View area} = 0.090 \times 0.060 = 0.0054\ \text{mm}^2$$

$$\text{Views on filter} = \frac{6.5^2 \times \pi}{0.0054} \cong 24{,}580$$

$$\text{Dilution factor} = 100$$

$$\text{Sample volume} = 0.02\ \text{mL}$$

$$\text{Average particle count} = 300$$

$$\text{Concentration} = \frac{300 \times 24{,}580}{0.02} \times 100 \cong 3.69 \times 10^{10}\ \text{p/mL}$$

The original concentration of the stock is 3.69×10^{10} particles/mL.

32. A collection of virus stocks can be combined at specific concentrations to produce an MC, for instance, with the same concentration for each virus. For spiking of fecal samples, the final concentration of the MC should be no less than 10^9 particles per gram of feces, to ensure sufficient sequencing depth of MC viruses.

3.2 Fecal Sample Preparation and Total Nucleic Acid Extraction

The following section of the protocol was in part adapted from the NetoVir protocol [27, 28], which is highly recommended.

1. Add 500 μL of sterile PBS to 50 mg of feces (fresh or from a frozen sample) in the ratio of 1:10 [w/v] (*see* **Note 7**).
2. Vortex for 1 min to homogenize the sample.
3. Centrifuge the samples at 17,000 × *g* for 3 min at 4 °C.
4. Transfer the supernatant containing virus-like particles, to a fresh 1.5-mL tube and centrifuge again at 17,000 × *g* for 3 min at 4 °C.
5. Filter the supernatant through a 0.8 μm PES membrane filter unit attached to a syringe into a sterile 1.5 mL tube.

6. Further, filter the solution from **step 5** through a 0.45 μm PES membrane filter unit into a sterile 1.5 mL tube.
7. Add 7 μL of 20× DNAse buffer (1 M Tris, 100 mM $CaCl_2$, and 30 mM $MGCl_2$, pH = 8) to 130 μL of filtrate.
8. Add 2 μL of TURBO DNAse and 13 μL of TURBO DNase 10× Reaction Buffer.
9. Incubate in a heat block for 2 h at 37 °C.
10. Stop the reaction by adding 1 μL 0.5 M EDTA (*see* **Note 8**).

    Optional: An aliquot of the suspension can be adsorbed onto a copper grid and negatively stained with 2% uranyl acetate to observe under transmission electron microscopy (TEM).
11. Nucleic acid extraction is performed using the Qiagen QIAamp Viral RNA Mini Kit without the inclusion of carrier RNA.
12. Add 140 μL of DNAse-treated filtrate to 560 μL AVL buffer in an RNAse-free 2 mL microcentrifuge tube and pulse vortex to mix.
13. Incubate at 20–22 °C for 10 min.
14. Briefly centrifuge to collect the sample at the bottom of the tube.
15. Add 560 μL absolute ethanol and mix by pulse vortex.
16. Briefly centrifuge to collect the sample at the bottom of the tube.
17. Add 630 μL of the sample mixture to the spin column.
18. Centrifuge at 6000 × *g* for 1 min.
19. Remove the column, add it to a fresh collection tube, and repeat **steps 17** and **18**.
20. Prepare the AW1 buffer according to Qiagen's guidelines.
21. Add 500 μL AW1 buffer to the spin column.
22. Centrifuge at 6000 × *g* for 1 min, remove the column, and add it to a fresh collection tube.
23. Prepare the AW2 buffer according to Qiagen's guidelines.
24. Add 500 μL AW2 buffer to the spin column.
25. Centrifuge at 17,000 × *g* for 3 min, remove the column, and add it to a fresh collection tube.
26. Centrifuge at 17,000 × *g* for 1 min, remove the column, and add it to a fresh RNAse-free 1.5 mL microcentrifuge tube.
27. Add 60 μL of AVE buffer to the spin column, and incubate at 20–22 °C for 1 min.
28. Centrifuge at 6000 × *g* for 1 min to elute the nucleic acids.

29. Store the eluted total nucleic acids at −80 °C, and store in aliquots where possible, to avoid freeze-thaw cycles during retrieval (*see* **Note 9**).

3.3 Reverse Transcription of RNA to cDNA

1. Conversion of RNA to cDNA is performed using the Sigma-Aldrich WTA2 Complete Whole Transcriptome Amplification Kit, modified as described below. Other than nuclease-free water, the reagents used in this section are taken from the WTA2 kit.
2. Aliquot 2.82 μL into a 0.2 mL PCR tube and add 0.5 μL Library Synthesis Solution. For large numbers of samples, 8-tube strips or a 96-well PCR plate may be more appropriate.
3. Mix by pipetting. If needed, briefly centrifuge the sample to collect the liquid.
4. Incubate in a thermocycler as follows:
   (a) 95 °C for 2 min
   (b) Hold at 18 °C.
5. To the cooled mixture, add 0.78 μL nuclease-free water, 0.5 μL Library Synthesis Buffer, and 0.4 μL Library Synthesis Enzyme. For multiple samples, prepare a master-mix and keep on ice for aliquoting into individual samples.
6. Mix by pipetting. If needed, briefly centrifuge the sample to collect liquid.
7. Incubate in a thermocycler as follows:
   (a) 18 °C for 10 min
   (b) 25 °C for 10 min
   (c) 37 °C for 30 min
   (d) 42 °C for 10 min
   (e) 70 °C for 20 min
   (f) Hold at 4 °C.
8. To the 5 μL DNA library that has now been prepared, add 60.2 μL nuclease-free water, 1.5 μL WTA2 dNTP mix, 7.5 μL Amplification Mix, and 0.75 μL Amplification Enzyme. If dealing with multiple samples, prepare a master-mix and keep it on ice for aliquoting into individual samples.
9. Incubate in a thermocycler as follows:
   (a) 94 °C for 2 min
   (b) 17 cycles of:
      (i) 94 °C for 30 s
      (ii) 70 °C for 5 min
   (c) Hold at 4 °C.

10. Purify the PCR product using a standard PCR purification kit such as the QIAquick PCR Purification Kit.
11. Quantify the amplified dsDNA using fluorimetry, such as the Qubit assay. Store at −80 °C prior to library preparation and sequencing.

3.4 Library Preparation and Sequencing

The mixture of DNA and second-strand synthesized cDNA generated in Subheading 3.3 can be used in traditional DNA sequencing methods, see Chaps. 7 and 8

3.5 Virome Bioinformatics Methods

3.5.1 Quality Control

The quality control of sequencing reads is a critical first step in bioinformatics analysis. While the main concepts and guidelines are covered in Chap. 12, this section outlines the essential quality control steps and details the unique challenges presented by virome datasets.

1. Assess quality of sequence reads with FastQC including base and sequence quality, GC content distribution, and presence of adapters or overrepresented sequences (*see* **Note 10**).
2. Document the initial read counts in the datasets before any filtering. Based on FastQC output, choose the quality filtering and trimming steps necessary for the dataset. The specific parameters chosen may be adjusted based on the sequencing platform used.
3. Perform quality control steps with Trimmomatic with these recommended parameters:
   (a) Minimum sequence read quality score (Phred score >30).
   (b) Adapter removal: Specify the adapter sequence appropriate for the library preparation.
   (c) Trimming of 5′ and/or 3′ ends based on per base quality scores and minimum read length threshold—typically 50 bp for short reads (*see* **Note 11**).
   (d) Remove reads with ambiguous bases (Ns).
4. Next, run FastQC again to verify the quality sequence improvement. Document the number of reads kept after quality control.
5. Remove host-derived reads by mapping against the appropriate reference genome (e.g., human or mouse genomes depending on sample origin) using bowtie2. Retain the unmapped reads for downstream analysis. Document the proportion of reads retained after this host decontamination step (*see* **Notes 12** and **13**).
6. Document the quality control parameter used and report the read loss at each step of the QC process. This allows the evaluation of potentially incorrect parameter choice and guarantees a high transparency of these preprocessing steps.

3.5.2 Assembly

Assembly of short reads (detailed in Chap. 8) aims to reconstruct viral genomes. Below are specific considerations to optimize assembly in virome datasets.

1. When analyzing time series or related samples (e.g., longitudinal sampling of the same individual), assemblies can be improved by grouping the reads of each individual sample (patient, animal, etc.) and performing a co-assembly.
2. Assemble the reads per library or per group with MEGAHIT [18] using:
   (a) --k-min 21 --k-max 149
   (b) --k-step 24
   (c) --min-contig-len 1500 (to reduce spurious viral contigs)
3. After assembly, evaluate the assembly quality by exploring the basic assembly statistics (N50, total assembly length, number, and length distribution of the contigs). To evaluate the proportion of reads omitted from the assembly process, consider mapping the reads back to the contigs and reporting the percentage of unmapped reads in each sample (*see* **Note 14**).
4. Post-assembly, different strategies may be applied to improve assembly quality:
   (a) Binning the viral contigs: vRhyme [29] supports viral sequence binning; however, improvements may be limited for "High-quality" contigs.
   (b) Phables [30, 31] can elevate some of the "Medium-quality" contigs to "High-quality" by refining the assembly graphs.

3.5.3 Mining Viral Contigs

Viral contig mining requires specialized tools due to the large viral diversity and rapid evolution of viral genomes. There are several dedicated tools for identifying viral contigs in assemblies, including VIBRANT [24], VirSorter2 [26], geNomad [14], and DeepVirFinder [11]. This section outlines a simple but robust approach to viral mining using complementary mining tools (*see* **Note 15**) to balance precision and recall of viral sequences.

1. Identify viral sequences using geNomad, and review the results of the summary/.virus_summary.tsv table. Retain only predictions on contigs longer than 1500 bp.
2. To increase the viral sequence recall, complementary mining using DeepVirFinder can be used. Retain prediction with a minimum p-value <0.05 and a score above 0.9 (*see* **Note 16**). While DeepVirFinder supports the mining of shorter viral sequences, we recommend keeping only sequences above 1500 bp.

3. For the integration of the results (*see* **Note 17**), create two viral sets: (1) a high-confidence set that contains viral sequences identified as viral by both tools and (2) a medium-confidence set containing contigs predicted only by one tool.

3.5.4 Filtering the Viral Genomes for High-quality Contigs Using CheckV

Quality assessment of the retrieved viral sequence is critical for downstream analysis and limiting false viral detection. CheckV provides standardized quality categories for viral sequence completeness and quality.

1. Estimate completeness of predicted viral genomes using CheckV [10] end-to-end command.
2. Parse checkV results from the quality_summary.tsv output table. Contigs will be given a percentage completeness estimate that corresponds to the following categories:
   (a) Complete (100% complete).
   (b) High-quality (≥90% complete).
   (c) Medium-quality (≥50–90% complete).
   (d) Low-quality (<50% complete).
   (e) Not determined (no completeness estimate).
3. Retain viral contigs with a minimum completeness score of 50% and higher for the downstream analyses.
4. Retain "Not determined" sequences with additional scrutiny, such as a length cut off (*see* **Note 18**) to limit the number of fragmented genomes (e.g., ≥10 kb) [32]. This collection may represent novel viruses distant from known references and can be considered for downstream analyses.
5. Record quality metrics by tracking the number of contigs in each category and reporting the sequence length distribution for each viral quality category.

3.5.5 Dereplication to vOTUs

To remove redundant viral contigs when using different miners, increase computational efficiency, and improve downstream read mapping results, it is recommended to cluster predicted viruses into viral operational taxonomic units (vOTUs) following the Minimum Information about an Uncultivated Virus Genome (MIUViG) guidelines [2]. This involves collapsing similar genomes into a single representative at the approximate level of species for tailed dsDNA bacterial and archaeal viruses (95% average nucleotide identity over 85% alignment fraction—relative to the shorter sequence). Use the anicalc and aniclust utilities provided by CheckV.

1. First, calculate the pairwise ANI: Create a blast database from the viral sequences obtained in Subheading 3.5.4 and do a self-blast of the viral sequences to themselves. Use the anicalc utility to obtain the pairwise ANI.

2. Use the aniclust utility script to parse the ANI results and generate the clusters using an ANI threshold of 95 and a coverage threshold of 85%.
3. Extract fasta sequences for cluster representatives using a sequence parsing tool, for example, Seqfu [33]. These representatives are the vOTUs (*see* **Note 19**).

3.5.6 Functional Annotation of vOTUs

The choice of annotation tools is dependent on the viral groups of interest (*see* **Note 20**). This section provides guidelines on bacteriophage sequences specifically.

1. Annotate the vOTUs with Pharokka to predict the CDS and assign function [19, 20, 34–45]. It is recommended to runn Pharokka with the following:
   (a) -g prodigal-gv to consider viruses that use alternative stop codons [46].
   (b) --meta and --split to annotate all vOTUs in one run. This speeds up run time when annotating large numbers of vOTUs.
2. The output files from Pharokka such as amino acid fasta files, annotated genbank files, and genomic fasta files are helpful for the downstream taxonomic assignments and other genome-level predictions.
3. Annotations can be supplemented with other tools, such as eggNOG mapper [47] and DRAM-v [48] to aid in the discovery of viral auxiliary metabolic genes.

3.5.7 Taxonomic Assignment

The taxonomic classification of viral contigs remains a challenging task requiring a combination of de novo clustering and machine-learning approaches to maximize the classification confidence. Tools like vConTACT2 [49] use shared protein clusters based on network-based approaches to link vOTUs to known viral clades offering insights into viral diversity and evolutionary relationships. GeNomad uses reference databases for assigning a broader genome and gene-level taxonomical classification although this is not the primary purpose of this tool.

1. Run vConTACT2 using the amino acid files from Pharokka and the gene of interest to the genome mapping file that links CDS to contigs. The vConTACT2 run can be performed with the built-in RefSeq database of bacteriophages, supplemented with the phage isolate-specific INPHARED database [34] that includes more reference genomes or with another viral reference database. If running vConTACT2 with INPHARED, the graphanalyzer.py [50] script can be used to assign taxonomy to vOTUs (*see* **Note 21**).

2. Run the geNomad [14] annotate pipeline with the dereplicated vOTU sequences as input for getting the broader taxonomy summary output, with classifications from realm to family, but not below (*see* **Note 22**).
3. Combine the information from both the tools for a consolidated taxonomy information of the vOTUs.

3.5.8 Handling Unclassified Viral Sequences

Many viral sequences may remain unclassified by the previous approach. This section provides strategies for a meaningful clustering of these unclassified viral sequences.

1. Many vOTUs in the realm *Duplodnaviria* may belong to unclassified orders and unclassified families within the class *Caudoviricetes.* Using the combined taxonomy information, group the vOTUs assigned as unclassified at the rank of realm with the *Duplodnaviria* (as they often lack the marker genes used to classify them as RNA and ssDNA viruses) to better understand the unclassified vOTUs from the realm to the family level. However, only do this for viromes where large numbers of tailed bacterial and/or archaeal viruses are expected to avoid data overinterpretation. Always document the classification rationale and the threshold used for reproducibility.
2. Cluster the grouped vOTUs based on their proteome similarity using the VipTreeGen server [25] or a standalone version and generate a dendrogram tree file (e.g., a Newick file) with branch lengths. Use appropriate reference viruses to contextualize.
3. Determine interim clades of vOTUs by splitting the tree at the branch length at which the known viral families diverge (~roughly at 0.4–0.5).
4. This interim placeholder with the clade names can be used in the downstream data analysis process to look at patterns (for example, longitudinal presence/absence of a specific interim clade across samples) if any instead of looking at them as broad "Unclassified" cluster.
5. If previously published catalogs are available for similar studies, classify the vOTUs if they are members of other metagenomic studies using specific metagenomic databases, for instance, the Unified Human Gut Virome Catalog for Human Gut Virome samples [51, 52]. Document any matches with these previously reported sequences.

3.5.9 Other Genome-Level Assignments

In this section, various tools are explored that can provide additional insights into the viral biology and ecology.

1. Host prediction: Run Integrated Prediction of Host-Phage interactions (iPHOP) with the dereplicated vOTUs list as

input to get host bacterium predictions for bacteriophages [53]. It is possible to leverage the taxonomy information to predict the host of certain viral groups, using the Virus Metadata Resource from the International Committee on Taxonomy of Viruses (available at https://ictv.global/vmr) (example: members of the family *Tombusviridae* are plant viruses).

2. Lifestyle prediction: the bacteriophage temperate or lytic lifestyle can be predicted (*see* **Note 23**) using the PhaTYP command from the PhaBOX toolkit 54.

3.5.10 vOTUs Abundance Estimation, Decontamination, and Denoising

To evaluate the relative abundance of the vOTU and viral groups identified above, an accurate abundance estimation in the sample is necessary.

1. Individually map the trimmed reads of each sample file against the vOTUs using a read aligner (e.g., bowtie2 or bwa) and samtools [8, 21] to produce a sorted bam file for each sample, including negative sequencing controls.
2. Use CoverM contig to produce a counts table of the samples using a normalized abundance metric. For viromes and for viruses predicted from bulk metagenomes, TPM (transcripts per million) and RPKM (reads per kilobase million), respectively are recommended. Set-min-read-percent-identity 90 and --min-read-aligned-percent 75 to reduce the impact of erroneous mapping [32].
3. Identify and remove contaminant and low-abundant vOTUs from the samples by carefully examining the vOTU abundance distribution of the negative control samples [55]. For example, remove any vOTUs from the analyses if they are highly abundant in the blank and samples above a certain threshold, as these vOTUs are likely to be introduced during sample processing. Low-abundant vOTUs are those that have TPM/RPKM values in all samples like the negative control and therefore do not give a signal above the noise threshold.
4. This vOTU table can be used to inform downstream alpha and beta diversity estimates, as described in Chap. 12.

4 Notes

1. Versions, tools, and programming languages listed are those used at the time of writing. Install the latest version and choice of each tool within a contained conda environment to avoid version conflicts and enhance reproducibility in the analysis.
2. Ideally, single EFM images contain between 200 and 600 particles. So, generally, between 3×10^6 and 1×10^7 particles

should be stained and filtered, depending on the size of the field of view in the microscope setup used. For this, a PFU count can be used as a starting point, but titration of the virus stock concentration is needed as the particle to PFU ratio of phages can be between 1:10 and 1:10,000 for eukaryotic virus stocks. Reliable stock concentrations can be determined by staining and imaging at least three replicates.

3. For a blank, dilution media can be used. Additionally, to control for possible host cell products, the filter-sterilized supernatant of a host cell stock can be used.
4. Particles can be counted manually, or image analysis software like ImageJ can be used to detect and count particles in the image.
5. Dimensions of the microscope view can typically be obtained from the microscope software or the image metadata.
6. As the outer edge of the Anodisc membrane is covered by the silicone gasket, it is recommended to determine the effective filtration area of the Anodisc membranes of the setup used. For instance, a fluorescently labelled high-concentration virus stock or bacterial stock can be used to measure the cross-section of the filtered area. In our experience, the cross-section of the filter is reduced by 1.6 mm, producing an effective filter area of 102.1 mm^2.
7. Include the mock community as a positive control and the buffer used for sample homogenization as a negative control throughout sample processing.
8. The volume of filtrate might vary from sample to sample. The viral-enriched samples can be stored at −80 °C until further processing although it is preferable to proceed directly to extraction as long-term storage may increase degradation of viral particles.
9. Take care to reduce the introduction of RNases: always wear gloves and use an RNase removal spray on surfaces and equipment.
10. An alternative tool for read QC is Cutadapt.
11. For paired-end datasets, ensure the proper handling of read pairs during the trimming and filtering steps.
12. High levels of host contamination (>90% of the quality-controlled reads mapped to the host genome) are expected for some sample types, such as skin or tissue samples. This can drastically affect the downstream analysis and should be taken into consideration in the experimental design.
13. An alternative tool for read mapping and host removal is bwa.

14. Alternative tool for assembly is metaSPAdes [22]. Previous research has shown [32, 56] that using metaSPAdes results in more high-quality genomes; however, the program uses more memory than MEGAHIT and might fail on larger datasets.
15. Use a combination of multiple viral mining tools that employ both Coding DNA Sequence (CDS) homology and sequence composition-based approaches and reconciling their outputs to enhance confidence in viral identification. The choice of miner(s) will be impacted by time and access to computational resources.
16. Miners are typically run with default parameters, but the resulting viral predictions may need to be filtered on miner-specific cut-offs including viral score and confidence values.
17. Tool selection depends on dataset features, computational resources, and training sets. Integrating results from multiple tools can improve sensitivity and precision but may increase complexity, and therefore, balance is the key [57–60].
18. Quality thresholds may require some adjustment according to the research aims. A strict filtering and retaining only "High-quality" sequences may be appropriate for genome-centered analysis while more permissive thresholds are appropriate for investigating the viral diversity in ecosystems.
19. CheckV completeness assessment (Subheading 3.5.5) may be performed after dereplication.
20. As Pharokka uses the PHROGs database [37] for annotation, it is well-suited for the annotation of bacteriophages but not for eukaryotic or archaeal viruses.
21. vConTACT2 clusters (VCs) represent viral groupings approximately to a level between genus and sub-family and do not necessarily correspond to actual taxa. Moreover, if expecting more eukaryotic viruses in the dataset, it is recommended to assign taxonomy using the NCBI Virus Database, by aligning nucleotide or protein sequences, for example, using BLAST or MMseqs2.
22. geNomad is not designed for taxonomic classification, and assignment should always be critically analyzed.
23. An alternative tool for bacteriophage lifestyle prediction is PHACTS [61].

Acknowledgements

This work was funded by the Biotechnology and Biological Sciences Research Council (BBSRC) Institute Strategic Programme Food Microbiome and Health BB/X011054/1 and its constituent projects; by the BBSRC Institute Strategic Programme Microbes and Food Safety BB/X011011/1 and its constituent projects; and by the BBSRC Core Capability Grant BB/CCG2260/1. R.C. and E.M.A. were also funded by a BBSRC grant BB/W015706/1. R.H. was supported by PhD studentships jointly funded by Invest in ME Research (UK Charity number 1153730) and the Faculty of Medicine and Health, University of East Anglia and a project grant awarded by ME Research UK (SCO Charity number SC036942) with the financial support of the Fred and Joan Davies bequest.

References

1. Adriaenssens EM, Roux S, Brister JR, Karsch-Mizrachi I, Kuhn JH, Varsani A, Yigang T, Reyes A, Lood C, Lefkowitz EJ, Sullivan MB, Edwards RA, Simmonds P, Rubino L, Sabanadzovic S, Krupovic M, Dutilh BE (2023) Guidelines for public database submission of uncultivated virus genome sequences for taxonomic classification. Nat Biotechnol 41:898–902. https://doi.org/10.1038/s41587-023-01844-2
2. Roux S, Adriaenssens EM, Dutilh BE, Koonin EV, Kropinski AM, Krupovic M, Kuhn JH, Lavigne R, Brister JR, Varsani A, Amid C, Aziz RK, Bordenstein SR, Bork P, Breitbart M, Cochrane GR, Daly RA, Desnues C, Duhaime MB, Emerson JB, Enault F, Fuhrman JA, Hingamp P, Hugenholtz P, Hurwitz BL, Ivanova NN, Labonté JM, Lee K-B, Malmstrom RR, Martinez-Garcia M, Mizrachi IK, Ogata H, Páez-Espino D, Petit M-A, Putonti C, Rattei T, Reyes A, Rodriguez-Valera F, Rosario K, Schriml L, Schulz F, Steward GF, Sullivan MB, Sunagawa S, Suttle CA, Temperton B, Tringe SG, Thurber RV, Webster NS, Whiteson KL, Wilhelm SW, Wommack KE, Woyke T, Wrighton KC, Yilmaz P, Yoshida T, Young MJ, Yutin N, Allen LZ, Kyrpides NC, Eloe-Fadrosh EA (2019) Minimum information about an uncultivated virus genome (MIUViG). Nat Biotechnol 37:29–37. https://doi.org/10.1038/nbt.4306
3. Hsieh S-Y, Tariq MA, Telatin A, Ansorge R, Adriaenssens EM, Savva GM, Booth C, Wileman T, Hoyles L, Carding SR (2021) Comparison of PCR versus PCR-free DNA library preparation for characterising the human faecal virome. Viruses 13:2093. https://doi.org/10.3390/v13102093
4. Trubl G, Solonenko N, Chittick L, Solonenko SA, Rich VI, Sullivan MB (2016) Optimization of viral resuspension methods for carbon-rich soils along a permafrost thaw gradient. PeerJ 4:e1999. https://doi.org/10.7717/peerj.1999
5. Bushnell B, Rood J, Singer E (2017) BBMerge—accurate paired shotgun read merging via overlap. PLoS One 12:e0185056. https://doi.org/10.1371/journal.pone.0185056
6. Quinlan AR, Hall IM (2010) BEDTools: a flexible suite of utilities for comparing genomic features. Bioinforma Oxf Engl 26:841–842. https://doi.org/10.1093/bioinformatics/btq033
7. Camacho C, Coulouris G, Avagyan V, Ma N, Papadopoulos J, Bealer K, Madden TL (2009) BLAST+: architecture and applications. BMC Bioinformatics 10:421. https://doi.org/10.1186/1471-2105-10-421
8. Langmead B, Salzberg SL (2012) Fast gapped-read alignment with bowtie 2. Nat Methods 9:357–359. https://doi.org/10.1038/nmeth.1923
9. Li H, Durbin R (2009) Fast and accurate short read alignment with Burrows–Wheeler transform. Bioinformatics 25:1754–1760. https://doi.org/10.1093/bioinformatics/btp324
10. Nayfach S, Camargo AP, Schulz F, Eloe-Fadrosh E, Roux S, Kyrpides NC (2021) CheckV assesses the quality and completeness of metagenome-assembled viral genomes. Nat Biotechnol 39:578–585. https://doi.org/10.1038/s41587-020-00774-7
11. Ren J, Song K, Deng C, Ahlgren NA, Fuhrman JA, Li Y, Xie X, Poplin R, Sun F (2020) Identifying viruses from metagenomic data using deep learning. Quant Biol 8:64–77. https://doi.org/10.1007/s40484-019-0187-4
12. Buchfink B, Reuter K, Drost H-G (2021) Sensitive protein alignments at tree-of-life scale using DIAMOND. Nat Methods 18:366–368. https://doi.org/10.1038/s41592-021-01101-x

13. Chen S, Zhou Y, Chen Y, Gu J (2018) Fastp: an ultra-fast all-in-one FASTQ preprocessor. Bioinformatics 34:i884–i890. https://doi.org/10.1093/bioinformatics/bty560
14. Camargo AP, Roux S, Schulz F, Babinski M, Xu Y, Hu B, Chain PSG, Nayfach S, Kyrpides NC (2024) Identification of mobile genetic elements with geNomad. Nat Biotechnol 42: 1303–1312. https://doi.org/10.1038/s41587-023-01953-y
15. Wickham H (2016) ggplot2. Springer, Cham
16. Steinegger M, Meier M, Mirdita M, Vöhringer H, Haunsberger SJ, Söding J (2019) HH-suite3 for fast remote homology detection and deep protein annotation. BMC Bioinformatics 20:473. https://doi.org/10.1186/s12859-019-3019-7
17. Letunic I, Bork P (2021) Interactive tree of life (iTOL) v5: an online tool for phylogenetic tree display and annotation. Nucleic Acids Res 49: W293–W296. https://doi.org/10.1093/nar/gkab301
18. Li D, Liu C-M, Luo R, Sadakane K, Lam T-W (2015) MEGAHIT: an ultra-fast single-node solution for large and complex metagenomics assembly via succinct de Bruijn graph. Bioinformatics 31:1674–1676. https://doi.org/10.1093/bioinformatics/btv033
19. Steinegger M, Söding J (2017) MMseqs2 enables sensitive protein sequence searching for the analysis of massive data sets. Nat Biotechnol 35:1026–1028. https://doi.org/10.1038/nbt.3988
20. Bouras G, Nepal R, Houtak G, Psaltis AJ, Wormald P-J, Vreugde S (2023) Pharokka: a fast scalable bacteriophage annotation tool. Bioinformatics 39:btac776. https://doi.org/10.1093/bioinformatics/btac776
21. Li H, Handsaker B, Wysoker A, Fennell T, Ruan J, Homer N, Marth G, Abecasis G, Durbin R, 1000 Genome Project Data Processing Subgroup (2009) The sequence alignment/map format and SAMtools. Bioinformatics 25:2078–2079. https://doi.org/10.1093/bioinformatics/btp352
22. Nurk S, Meleshko D, Korobeynikov A, Pevzner PA (2017) metaSPAdes: a new versatile metagenomic assembler. Genome Res 27: 824–834. https://doi.org/10.1101/gr.213959.116
23. Bolger AM, Lohse M, Usadel B (2014) Trimmomatic: a flexible trimmer for Illumina sequence data. Bioinformatics 30:2114–2120. https://doi.org/10.1093/bioinformatics/btu170
24. Kieft K, Zhou Z, Anantharaman K (2020) VIBRANT: automated recovery, annotation and curation of microbial viruses, and evaluation of viral community function from genomic sequences. Microbiome 8:90. https://doi.org/10.1186/s40168-020-00867-0
25. Nishimura Y, Yoshida T, Kuronishi M, Uehara H, Ogata H, Goto S (2017) ViPTree: the viral proteomic tree server. Bioinformatics 33:2379–2380. https://doi.org/10.1093/bioinformatics/btx157
26. Guo J, Bolduc B, Zayed AA, Varsani A, Dominguez-Huerta G, Delmont TO, Pratama AA, Gazitúa MC, Vik D, Sullivan MB, Roux S (2021) VirSorter2: a multi-classifier, expert-guided approach to detect diverse DNA and RNA viruses. Microbiome 9:37. https://doi.org/10.1186/s40168-020-00990-y
27. Conceição-Neto N, Zeller M, Lefrère H, De Bruyn P, Beller L, Deboutte W, Yinda CK, Lavigne R, Maes P, Ranst MV, Heylen E, Matthijnssens J (2015) Modular approach to customise sample preparation procedures for viral metagenomics: a reproducible protocol for virome analysis. Sci Rep 5:16532. https://doi.org/10.1038/srep16532
28. Conceição-Neto N, Yinda KC, Van Ranst M, Matthijnssens J (2018) NetoVIR: modular approach to customize sample preparation procedures for viral metagenomics. Methods Mol Biol Clifton NJ 1838:85–95. https://doi.org/10.1007/978-1-4939-8682-8_7
29. Kieft K, Adams A, Salamzade R, Kalan L, Anantharaman K (2022) vRhyme enables binning of viral genomes from metagenomes. Nucleic Acids Res 50:e83. https://doi.org/10.1093/nar/gkac341
30. Mallawaarachchi V, Roach MJ, Decewicz P, Papudeshi B, Giles SK, Grigson SR, Bouras G, Hesse RD, Inglis LK, Hutton ALK, Dinsdale EA, Edwards RA (2023) Phables: from fragmented assemblies to high-quality bacteriophage genomes. Bioinformatics 39:btad586. https://doi.org/10.1093/bioinformatics/btad586
31. Cook R, Telatin A, Hsieh S-Y, Newberry F, Tariq MA, Baker DJ, Carding SR, Adriaenssens EM (2024) Nanopore and Illumina sequencing reveal different viral populations from human gut samples. Microb Genom 10: 001236. https://doi.org/10.1099/mgen.0.001236
32. Roux S, Emerson JB, Eloe-Fadrosh EA, Sullivan MB (2017) Benchmarking viromics: an in silico evaluation of metagenome-enabled estimates of viral community composition and diversity. PeerJ 5:e3817. https://doi.org/10.7717/peerj.3817
33. Telatin A, Fariselli P, Birolo G (2021) SeqFu: a suite of utilities for the robust and reproducible

manipulation of sequence files. Bioengineering 8:59. https://doi.org/10.3390/bioengineering8050059

34. Cook R, Brown N, Redgwell T, Rihtman B, Barnes M, Clokie M, Stekel DJ, Hobman J, Jones MA, Millard A (2021) INfrastructure for a PHAge REference database: identification of large-scale biases in the current collection of cultured phage genomes. PHAGE 2:214–223. https://doi.org/10.1089/phage.2021.0007
35. Chan PP, Lin BY, Mak AJ, Lowe TM (2021) tRNAscan-SE 2.0: improved detection and functional classification of transfer RNA genes. Nucleic Acids Res 49:9077–9096. https://doi.org/10.1093/nar/gkab688
36. Ondov BD, Treangen TJ, Melsted P, Mallonee AB, Bergman NH, Koren S, Phillippy AM (2016) Mash: fast genome and metagenome distance estimation using MinHash. Genome Biol 17:132. https://doi.org/10.1186/s13059-016-0997-x
37. Terzian P, Olo Ndela E, Galiez C, Lossouarn J, Pérez Bucio RE, Mom R, Toussaint A, Petit M-A, Enault F (2021) PHROG: families of prokaryotic virus proteins clustered using remote homology. NAR Genom Bioinforma 3:lqab067. https://doi.org/10.1093/nargab/lqab067
38. Bland C, Ramsey TL, Sabree F, Lowe M, Brown K, Kyrpides NC, Hugenholtz P (2007) CRISPR recognition tool (CRT): a tool for automatic detection of clustered regularly interspaced palindromic repeats. BMC Bioinformatics 8:209. https://doi.org/10.1186/1471-2105-8-209
39. Laslett D, Canback B (2004) ARAGORN, a program to detect tRNA genes and tmRNA genes in nucleotide sequences. Nucleic Acids Res 32:11–16. https://doi.org/10.1093/nar/gkh152
40. Chen L, Yang J, Yu J, Yao Z, Sun L, Shen Y, Jin Q (2005) VFDB: a reference database for bacterial virulence factors. Nucleic Acids Res 33: D325–D328. https://doi.org/10.1093/nar/gki008
41. Alcock BP, Raphenya AR, Lau TTY, Tsang KK, Bouchard M, Edalatmand A, Huynh W, Nguyen A-LV, Cheng AA, Liu S, Min SY, Miroshnichenko A, Tran H-K, Werfalli RE, Nasir JA, Oloni M, Speicher DJ, Florescu A, Singh B, Faltyn M, Hernandez-Koutoucheva A, Sharma AN, Bordeleau E, Pawlowski AC, Zubyk HL, Dooley D, Griffiths E, Maguire F, Winsor GL, Beiko RG, Brinkman FSL, Hsiao WWL, Domselaar GV, McArthur AG (2020) CARD 2020: antibiotic resistome surveillance with the comprehensive antibiotic resistance database. Nucleic Acids Res 48:D517–D525. https://doi.org/10.1093/nar/gkz935
42. Larralde M (2022) Pyrodigal: python bindings and interface to prodigal, an efficient method for gene prediction in prokaryotes. J Open Sour Softw 7:4296. https://doi.org/10.21105/joss.04296
43. Soft LM, Zeller G (2023) PyHMMER: a python library binding to HMMER for efficient sequence analysis. Bioinformatics 39: btad214. https://doi.org/10.1093/bioinformatics/btad214
44. Larralde M, Camargo A (2023) Pyrodigal-gv: a pyrodigal extension to predict genes in giant viruses and viruses with alternative genetic code. https://github.com/althonos/pyrodigal-gv
45. Shimoyama, Y. (2022). pyCirclize: Circular visualization in Python [Computer software]. https://github.com/moshi4/pyCirclize date-released 12–20
46. Cook R, Telatin A, Bouras G, Camargo AP, Larralde M, Edwards RA, Adriaenssens EM (2024) Driving through stop signs: predicting stop codon reassignment improves functional annotation of bacteriophages. ISME Commun 4:ycae079. https://doi.org/10.1093/ismeco/ycae079
47. Cantalapiedra CP, Hernández-Plaza A, Letunic I, Bork P, Huerta-Cepas J (2021) eggNOG-mapper v2: functional annotation, orthology assignments, and domain prediction at the metagenomic scale. Mol Biol Evol 38: 5825–5829. https://doi.org/10.1093/molbev/msab293
48. Shaffer M, Borton MA, McGivern BB, Zayed AA, La Rosa SL, Solden LM, Liu P, Narrowe AB, Rodríguez-Ramos J, Bolduc B, Gazitúa MC, Daly RA, Smith GJ, Vik DR, Pope PB, Sullivan MB, Roux S, Wrighton KC (2020) DRAM for distilling microbial metabolism to automate the curation of microbiome function. Nucleic Acids Res 48:8883–8900. https://doi.org/10.1093/nar/gkaa621
49. Ho Bin Jang H, Bolduc B, Zablocki O, Kuhn JH, Roux S, Adriaenssens EM, Brister JR, Kropinski AM, Krupovic M, Lavigne R, Turner D, Sullivan MB (2019) Taxonomic assignment of uncultivated prokaryotic virus genomes is enabled by gene-sharing networks. Nat Biotechnol 37:632–639. https://doi.org/10.1038/s41587-019-0100-8
50. Pandolfo M, Telatin A, Lazzari G, Adriaenssens EM, Vitulo N (2022) MetaPhage: an automated pipeline for analyzing, annotating, and classifying bacteriophages in metagenomics sequencing data. mSystems 7:e0074122.

https://doi.org/10.1128/msystems.00741-22

51. Almeida A, Nayfach S, Boland M, Strozzi F, Beracochea M, Shi ZJ, Pollard KS, Sakharova E, Parks DH, Hugenholtz P, Segata N, Kyrpides NC, Finn RD (2021) A unified catalog of 204,938 reference genomes from the human gut microbiome. Nat Biotechnol 39:105–114. https://doi.org/10.1038/s41587-020-0603-3
52. Galperina A, Lugli GA, Milani C, Vos WMD, Ventura M, Salonen A, Hurwitz B, Ponsero AJ (2024) The aggregated gut viral catalogue (AVrC): a unified resource for exploring the viral diversity of the human gut. 2024.06.24.600367
53. Roux S, Camargo AP, Coutinho FH, Dabdoub SM, Dutilh BE, Nayfach S, Tritt A (2023) iPHoP: an integrated machine learning framework to maximize host prediction for metagenome-derived viruses of archaea and bacteria. PLoS Biol 21:e3002083. https://doi.org/10.1371/journal.pbio.3002083
54. Shang J, Tang X, Sun Y (2023) PhaTYP: predicting the lifestyle for bacteriophages using BERT. Brief Bioinform 24:bbac487. https://doi.org/10.1093/bib/bbac487
55. Kuzub N, Kurilshikov A, Zhernakova A, Garmaeva S (2024) Negativeome in early-life virome studies: characterization and decontamination. 2024.10.14.618243
56. Sutton TDS, Clooney AG, Ryan FJ, Ross RP, Hill C (2019) Choice of assembly software has a critical impact on virome characterisation. Microbiome 7:12. https://doi.org/10.1186/s40168-019-0626-5
57. Hegarty B, Riddell VJ, Bastien E, Langenfeld K, Lindback M, Saini JS, Wing A, Zhang J, Duhaime M (2024) Benchmarking informatics approaches for virus discovery: caution is needed when combining in silico identification methods. mSystems 9:e01105-23. https://doi.org/10.1128/msystems.01105-23
58. Ho SFS, Wheeler NE, Millard AD, van Schaik W (2023) Gauge your phage: benchmarking of bacteriophage identification tools in metagenomic sequencing data. Microbiome 11:84. https://doi.org/10.1186/s40168-023-01533-x
59. Schackart KE, Graham JB, Ponsero AJ, Hurwitz BL (2023) Evaluation of computational phage detection tools for metagenomic datasets. Front Microbiol 14. https://doi.org/10.3389/fmicb.2023.1078760
60. Wu X, Zhu J, Tao P, Rao VB (2021) Bacteriophage T4 escapes CRISPR attack by minihomology recombination and repair. mBio 12:e0136121. https://doi.org/10.1128/mBio.01361-21
61. McNair K, Bailey BA, Edwards RA (2012) PHACTS, a computational approach to classifying the lifestyle of phages. Bioinforma Oxf Engl 28:614–618. https://doi.org/10.1093/bioinformatics/bts014

Chapter 15

Mycobiomics

Steve James and Ezgi Özkurt

Abstract

The ability to define and characterize the variety of unicellular yeasts and filamentous fungi that inhabit the human gastrointestinal (GI) tract is hampered by the lack of suitable methodologies. This chapter describes culture-based protocols to isolate and culture viable enteric fungi from fresh or refrigerated human fecal samples. It also describes non-culture-based protocols to isolate fungal DNA from human fecal samples to use as a template for fungal community profiling using a high-throughput internal transcribed spacer region 1 (ITS1) sequencing approach. By addressing the major challenges specific to ITS datasets and presenting methods to circumvent them, researchers can select appropriate approaches to ensure robust and meaningful ITS data interpretations.

Key words Fungi, Mycobiome, Gastrointestinal tract, Internal transcribed spacer, Amplicon sequencing, Data analysis

1 Introduction

The human gut harbors a diverse variety of unicellular yeasts and filamentous fungi, predominantly from the Ascomycota phylum [1–3], with fungi typically comprising less than 0.1% of the total microbiota in the healthy adult gut [4, 5]. However, fungi are larger than bacteria by at least an order of magnitude [6], meaning they constitute a sizeable fraction of the microbial biomass [3], contributing to human health, maintenance of intestinal homeostasis, and development of the host immune system [1, 7, 8]. An imbalance in mycobiota composition has been implicated in various diseases [9–15].

Important considerations are the inclusion of antibiotic-supplemented selective growth media to suppress/inhibit bacterial growth and a combination of chemical and mechanical lysis approaches to maximize DNA recovery. Amplicon-based sequencing can help address the low abundance of fungi in biological samples such as feces. For instance, fungi typically have more copies of the ribosomal rRNA genes that collectively make up the

Simon R. Carding (ed.), *Best Practice in Microbiome Research*, Springer Protocols Handbooks,
https://doi.org/10.1007/978-1-0716-5009-7_15,

ribosomal DNA (rDNA) array, than their bacterial counterparts. For example, *Saccharomyces cerevisiae*, a food-associated yeast often found in the human gut, can have between 50 and 100 copies per haploid genome depending on the strain [16]. Currently, the multi-copy non-coding ITS1 and ITS2 regions of the rDNA array are widely used as taxonomic biomarkers (DNA barcodes) for high-throughput NGS-based amplicon sequencing and fungal community profiling [2, 17, 18].

2 Materials

2.1 Fungal Cultivation

1. Sterile 1× PBS buffer (used as a diluent).
2. SD broth (40 g/L dextrose and 10 g/L peptone) supplemented with chloramphenicol (0.05 g/L) and kanamycin (0.05 g/L).
3. SD agar (40 g/L dextrose, 10 g/L peptone, and 20 g/L agar) supplemented with chloramphenicol (0.05 g/L) and kanamycin (0.05 g/L).
4. YM broth (10 g/L glucose, 3 g/L malt extract, 5 g/L peptone, and 3 g/L yeast extract) supplemented with penicillin (25 U/mL) and streptomycin (25 U/mL).
5. YM agar (10 g/L glucose, 3 g/L malt extract, 5 g/L peptone, 3 g/L yeast extract, and 20 g/L agar) supplemented with penicillin (25 U/mL) and streptomycin (25 U/mL).
6. Petri dishes (triple vent).
7. 20 mL Glass bottles (plastic capped).
8. 15 mL Corning® centrifuge tubes.
9. Sterile spatulas.
10. Benchtop vortex (e.g., Vortex-Genie®).
11. Sterile, aerosol-resistant pipette tips (of varying volume capacities: 1 μL to 1 mL).
12. Plastic pellet pestles (sterile).
13. Plastic L-shaped spreaders (sterile).
14. Parafilm.
15. Incubator (set to 30 °C).
16. Sterile 20% glycerol.
17. Bioguard disinfectant surface spray.

2.2 Non-culture-Based Materials

1. Benchtop microcentrifuge (up to 16,000×*g*).
2. 1.5 mL LoBind microcentrifuge tubes.
3. QIAamp PowerFecal Pro DNA Kit (Qiagen).

4. FastPrep-24 benchtop homogeniser (MP Biomedicals).
5. Qubit 4.0 Fluorometer (Invitrogen).
6. Qubit dsDNA Broad-Range (BR) assay kit (Invitrogen).
7. NanoDrop spectrophotometer.
8. 0.6 mL microcentrifuge tubes.
9. 0.2 mL PCR tube strips and lids.
10. KAPA2G Robust PCR kit (Roche).
11. Mini-fuge with 8-place rotor for 0.2-mL tubes.
12. Thermal cycler, 0.2-mL tube compatible (e.g., Biometra TRIO instrument).
13. Midori Green Direct DNA stain.
14. Blue/Green LED Transilluminator.
15. KAPA Pure beads (Roche).
16. Elution buffer (EB) (10 mM Tris-HCl).

3 Methods

3.1 Culture-Based Protocol

Selective media supplemented with antibiotics should be used for the cultivation of viable enteric fungi from fecal samples (*see* **Note 1**).

3.1.1 Fecal Homogenate Preparation

Fresh or refrigerated human fecal samples should be used as freezing samples prior to culturing can significantly reduce the recovery of viable fungi. A single freeze/thaw cycle can result in a tenfold reduction in fungal abundance compared with fresh samples [19]. All processing steps should be carried out in a Class 2 Safety Cabinet, which should be sterilized prior to use (e.g., Bioguard disinfectant surface spray).

1. Using a sterile spatula, weigh out between 0.1 and 1.0 g of feces into a sterile 15 mL Corning® centrifuge tube or similar (*see* **Note 2**).
2. Add sterile 1× PBS buffer to the centrifuge tube to produce an initial 10× diluted suspension (e.g., for 0.5 g of feces, add 5.0 mL 1× PBS).
3. Vortex at maximum speed for 30–60 s to produce a uniform homogenate. If necessary, a sterile plastic pellet pestle or 1 mL pipette tip can be used to break up any large particulate matter that may remain in suspension (after the initial vortex), before a second vortex. The sample is now ready to use for culturing.

3.1.2 Semi-anaerobic Liquid Culturing

1. Add 200 μL of fecal homogenate to 10 mL of SD broth (supplemented with chloramphenicol and kanamycin) pre-dispensed in a sterile 20 mL glass universal bottle, and vortex thoroughly to mix. A duplicate culture should be set up if there is sufficient homogenate (*see* **Note 3**).
2. Repeat **step 1** but add 200 μL of homogenate to 10 mL of YM broth (supplemented with penicillin and streptomycin). Again, set up in duplicate if sufficient homogenate is available.
3. Incubate all broth cultures at 30 °C without agitation in the dark.
4. Broth cultures should be monitored daily for signs of microbial growth indicated by an increase in turbidity, appearance of gas bubbles or both.
5. Use standard light microscopy to monitor fungal growth using a 3–4 μL aliquot of culture medium examined at 200× and 400× magnification. Yeast and other fungal cells can be readily distinguished from bacterial cells based on their larger size [20] (*see* **Note 4**).
6. Once fungal growth is detected, each liquid culture is serially diluted in sterile 1× PBS buffer and spread plated onto agar plates of the same selective medium type (e.g., YM broth onto YM agar).

3.1.3 Aerobic Culturing

1. For aerobic cultivation by spread plating, first set up a series of tenfold dilutions from the fecal homogenate, and the homogenate-derived SD and YM liquid cultures (dilution range: 10^{-1} to 10^{-4}).
2. To set up the first tenfold dilution (i.e., 10^{-1} or 10× dilution), add 100 μL of the fecal homogenate (or liquid culture) to a sterile 1.5 mL centrifuge tube containing 900 μL sterile PBS (diluent). Vortex to mix thoroughly.
3. Carry out a second tenfold dilution by transferring 100 μL of the first tenfold dilution (**step 2**) into a new sterile 1.5 mL centrifuge tube containing 900 μL sterile PBS and use a vortex to mix. This is the 10^{-2} or 100× dilution.
4. Repeat the process twice more to produce a total of 4 tenfold dilutions, from 10^{-1} to 10^{-4} (i.e., from 10× to 10,000× diluted, respectively).
5. For cultivation on solid medium, pipette a 50 μL aliquot of each dilution (10^{-1} to 10^{-4}) onto a separate antibiotic-supplemented agar plate (either SD or YM).
6. For each plate, use a separate sterile plastic L-shaped spreader, to spread the liquid evenly across the entire surface of the agar plate, carefully rotating the Petri dish at the same time. Allow each agar plate to air dry for 5 min at 15–25 °C.

7. For the fecal homogenate, two sets of plates should be set up. One set of antibiotic-supplemented SD plates (Set #1) and one set of antibiotic-supplemented YM plates (Set #2). In the case of the SD and YM liquid cultures, each dilution series need only be plated onto antibiotic-supplemented agar of the same medium type.
8. A 50 μL aliquot of diluent (i.e., PBS) should be spread plated onto a separate agar plate to act as a negative control and to monitor for microbial contamination. Likewise, a positive growth control should be included, e.g., *Candida albicans*. Allow each plate to air dry for 5 min at 15–25 °C.
9. Once dry, transfer all plates to a 30 °C incubator, and incubate inverted in the dark.
10. After 2 days incubation, inspect all plates for signs of fungal growth, notably for fast-growing species (e.g., *C. albicans*).
11. Continue incubating all plates for a further 5–7 days to allow detection and cultivation of slower-growing fungi.
12. Pick at least two well-separated colonies of each unique colony type (i.e., morphotype) from each spread plate showing microbial growth, and re-plate onto fresh agar of the same medium type. There should be no need to use antibiotic-supplemented medium, but it can be used if bacterial growth persists (*see* **Note 5**).
13. Incubate plates inverted at 30 °C in the dark (*see* **Note 6**).
14. Once good growth has been achieved, a sample of each colony type can be picked and stored (banked) frozen at −70 °C in sterile 20% glycerol stock, prior to further characterization.

3.2 Non-culture-Based Protocol

3.2.1 DNA Extraction

The following protocol uses the commercial Qiagen QIAamp PowerFecal Pro DNA Kit in combination with high-speed bead beating, which is an effective manual method for maximizing the recovery of low-abundance fungal DNA from human fecal samples (*see* **Note 7**).

Note that all centrifugation steps should be performed at 15–25 °C.

1. Use a sterile spatula to add 200–250 mg of feces to a clean 1.5 mL microcentrifuge tube.
2. Add 800 μL of solution CD1 and vortex vigorously to mix. A plastic pellet pestle can be used to break up any remaining large particulate matter. Repeat if necessary to produce a uniform homogenate.
3. Transfer the homogenate to a PowerBead Pro tube using a 1 mL pipette tip (*see* **Note 8**).

4. Homogenize samples using a FastPrep-24 benchtop instrument (MP Biomedicals) set at 6.0 m/s for 1 min. If larger-sized DNA of higher molecular weight is required (e.g., for ONT long-read sequencing), vortex to homogenize using a benchtop vortex at maximum speed for 10 min. Tubes should be secured horizontally on a vortex adapter for 1.5–2 mL tubes.
5. Centrifuge the PowerBead Pro tube at 15,000×*g* for 1 min.
6. Transfer the supernatant to a clean 2 mL microcentrifuge tube.
7. Add 200 μL of solution CD2 and vortex for 5 s.
8. Centrifuge at 15,000×*g* for 1 min. Transfer the supernatant (up to 700 μL) to a clean 2 mL microcentrifuge tube.
9. Add 600 μL of solution CD3 and vortex for 5 s.
10. Load 650 μL of the lysate onto an MB Spin Column (supplied in kit) and centrifuge at 15,000×*g* for 1 min.
11. Discard the flow-through and repeat **step 10** to ensure all lysate has passed through the spin column.
12. Carefully place the spin column into a clean 2 mL microcentrifuge collection tube avoiding transfer of any flow-through onto the spin column.
13. Add 500 μL of solution EA to the spin column and centrifuge at 15,000×*g* for 1 min.
14. Discard the flow-through and place the spin column back into the same collection tube.
15. Add 500 μL of solution C5 to the spin column and centrifuge at 15,000×*g* for 1 min.
16. Discard the flow-through and place the spin column into a new 2 mL collection tube and centrifuge at 16,000×*g* for 2 min.
17. Place the spin column into a new 1.5 mL elution tube. To maximize DNA recovery, use 1.5 mL LoBind microcentrifuge tubes in place of the elution tubes supplied.
18. Add 50 μL of solution C6 (10 mM Tris-HCl) to the center of the filter membrane.

    Do not disturb the (white) filter membrane when pipetting.
19. Centrifuge at 15,000×*g* for 1 min. Discard the spin column. Solution C6 does not contain EDTA, and so the DNA should be stored at −70 °C or below.

3.2.2 Quantification and Quality Checking

The Qubit 4.0 Fluorometer and associated Qubit dsDNA Broad-Range (BR) assay kit (quantification range: 1–1000 ng/μL) are suitable for routine DNA quantification and quality checking as follows:

1. Set up two assay tubes for the DNA standards (Standard #1 and Standard #2) and one assay tube for each DNA sample.
2. Prepare the Qubit working solution by diluting the Qubit reagent concentrate 1:200 in Qubit diluent buffer. Prepare 200 μL of working solution for each standard and DNA sample.
3. For each standard, add 190 μL working solution to the 10 μL standard. For each DNA sample, add 198 to 2 μL DNA.
4. Vortex all tubes for 2–3 s and incubate for 2 min at 15–25 °C.
5. As prompted, set up a calibration curve by inserting each standard (Standard #1 first, and Standard #2 second) into the Qubit 4.0 Fluorometer and taking a reading.
6. Once the calibration curve is set up, insert each DNA sample separately into the Qubit 4.0 Fluorometer and take a reading (*see* **Note 9**).

Further details on general DNA QC are provided in Chap. 5.

3.2.3 Amplification of the Fungal ITS1 Region

The following protocol can be used to routinely prepare fungal ITS1 amplicon libraries for high-throughput sequencing using the Illumina NextSeq 2000 instrument. For amplification of the ITS1 region, use the "universal" fungal primers ITS1F and ITS2 [21, 22], modified to include the Illumina sequencing adapters, thus allowing a one-step amplification prior to indexing PCR. The Illumina-modified ITS1 primer sequences are shown below (5′ → 3′), with the adapter sequences underlined:

ITS1F—TCGTCGGCAGCGTCAGATGTGTATAAGAGACAGC TTGGTCATTTAGAGGAAGTAA

ITS2—GTCTCGTGGGCTCGGAGATGTGTATAAGAGACAG GCTGCGTTCTTCATCGATGC

The ITS1 amplicon library PCR is detailed below. All reagents, as well as template DNA, once thawed, should be kept on ice during PCR setup. The setup time should be kept as short as possible to prevent reagents and DNA from warming to ambient temperature. All amplifications are carried out in a 25 μL reaction volume in 0.2 mL PCR tube strips (*see* **Note 10**).

1. PCR master mix: add the following reagents to a 0.6 mL microcentrifuge tube in the order shown below:

PCR-grade H_2O	8.9 μL
5× GC buffer (included in kit)[a]	5.0 μL
Primer mix (10 pmol/μL)	0.5 μL
dNTP mix (10 mM each)	0.5 μL
KAPA2G DNA polymerase (5 U/μL)	0.1 μL

[a]Contains 1.5 mM $MgCl_2$ at 1× concentration

All volumes are for a single reaction and do not include the template. When preparing the PCR master mix, prepare for $n + 1$ samples to allow for possible pipetting error.

2. Vortex gently to mix and briefly centrifuge to ensure all contents collect at the bottom of the tube.
3. Place the PCR master mix on ice.
4. Add 10 μL of template DNA (10 ng/μL) to each individual 0.2 mL tube. Duplicate amplification reactions should be set up for each fecal DNA sample (*see* **Note 11**).
5. Positive and negative controls should be included in each separate PCR run. As a positive control, use 0.1 ng of *C. albicans* genomic DNA, and PCR-grade water as a negative control.
6. Once all DNA templates and controls have been added, briefly centrifuge the PCR tube strips to ensure all contents collect at the bottom of the tubes. A mini-fuge with an 8-place rotor for 0.2 μL tubes can be used for this purpose.
7. Add 15 μL of PCR master mix to the top of each tube.
8. Add lid strips, ensuring that each individual lid is firmly pressed into place.
9. Briefly centrifuge the PCR tube strips to ensure all contents collect at the bottom of the tubes.
10. Load the PCR tube strips onto a 0.2 μL tube compatible thermal cycler with a heated lid (e.g., Biometra TRIO).
11. For fungal ITS1 amplification, when using either standard or modified ITS1F and ITS2 primers, use the following cycling parameters:

1 cycle	95 °C, 5 min
35 cycles	95 °C, 30 s; 55 °C, 30 s; 72 °C, 30 s
1 cycle	72 °C, 5 min

The temperature of the heated lid should be set to 99 °C to prevent sample evaporation.

12. Use 1% agarose gel electrophoresis to confirm PCR amplification and provide an estimate of the ITS1 amplicon size. The size of the ITS1 region can vary between different fungal species and genera, typically ranging from 100 to 400 bp in length [23, 24] (*see* **Note 12**).
13. Add 0.5 μL Midori Green Direct DNA stain to a 5 μL aliquot of each PCR sample (*see* **Note 13**). Vortex to mix.

14. Load each sample onto a 1% agarose gel, along with a 100-bp DNA ladder, and run at 100 V until the dye front has migrated at least halfway down the gel.
15. Visualize gel under blue light using a blue/green LED transilluminator.

3.2.4 Library Preparation and Sequencing

Following agarose gel electrophoresiss, sample replicates are combined and a 0.7× SPRI purification is performed using KAPA Pure Beads with the purified DNA eluted in 20 μL of EB buffer (10 mM Tris-HCl). The purified DNA is then ready for the second PCR (i.e., the Indexing PCR), where the sequencing barcodes are added to each sample (see Chaps. 6 and 7 for details). After a second SPRI clean-up, samples are quantified, normalized, and pooled in equal quantities. ITS1 library pools are run at a final concentration of 8 pM, on an Illumina NextSeq2000 instrument using the Illumina P1 Reagent Kit (600-cycle) with 2× 300 bp output (see Chaps. 6 and 8).

3.3 ITS Amplicon Data Analysis

This section provides recommendations on the workflow of ITS amplicon data analysis, exploring the available pipelines and discussing critical considerations for achieving accurate quality filtering, clustering, taxonomic assignments, and phylogenetic insights.

3.3.1 Pipelines for ITS Data Analysis

ITS data analysis is similar to 16S rRNA data analysis (see Chap. 6). However, the ITS region possesses unique characteristics that require special attention during data processing. These include higher intra-specific variation [25], significant length variation, frequent alignment gaps [24], and homopolymers [26]. To address these challenges, tailored processing approaches are needed, choosing between:

1. Multi-marker pipelines, such as LotuS2 [27], QIIME2 [28], PipeCraft2 [29], and AMPtk [30]. These are versatile, support the processing of ITS datasets, offer features tailored for ITS analysis, and include functionality for extracting the ITS region from sequence reads. For example, LotuS2 [27] provides configuration files where defaults are provided for ITS datasets, while AMPtk [30] is optimized to handle variable-length sequences, a common feature of fungal ITS datasets.
2. Pipelines such as FROGS [31], DAnIEL [32], and PIPITS [33], which are exclusively designed for analyzing ITS datasets.

3.3.2 Dataset Filtering

The fungal ITS region ranges from 50 bases to several kilobases in length, [24] which presents challenges during the initial analysis steps, as datasets may contain non-overlapping reads due to the presence of long ITS sequences [34]. Merging forward and reverse reads from such regions in paired-end sequencing data, often fails, leading to these reads being filtered out. However, exclusion of these non-overlapping reads can result in the systematic removal of taxa with longer ITS regions, introduce bias, and provide an incomplete representation of fungal diversity which can be addressed by considering:

1. Some pipelines (e.g., FROGS [31], PipeCraft2 [29], Dadaist2 [35], and Cascabel [36]) include settings that accommodate both overlapping and non-overlapping reads, to ensure a more comprehensive and accurate characterization of fungal composition.
2. The fungal ITS region often contains regions of repetitive DNA sequence (homopolymers), which can lead to sequencing errors and hamper sequence alignment. To overcome this, some pipelines include a default filtering step to remove them (e.g., QIIME2). While useful for filtering 16S rRNA sequences, this step is unsuitable for ITS sequences, which can often contain homopolymeric regions of ten bases or more. Therefore, homopolymer filtering should be avoided when analyzing ITS datasets [26].
3. In fungi, the ITS region is located between the small- (18S) and large-subunit (28S) rRNA genes, separated into two hypervariable sub-regions (ITS1 and ITS2), by the intervening 5.8S rRNA gene [37]. Amplification of either the ITS1 or ITS2 region uses primers targeting the conserved 18S/5.8S or 5.8S/28S rRNA regions, respectively [24]. Consequently, removal of conserved flanking region sequences (i.e., 18S, 5.8S, and 28S) is a crucial filtering step in ITS sequence extraction and for downstream analysis [38]. It avoids inflated biodiversity estimates, improves sequence clustering, and ensures optimal taxonomic assignments (*see* **Note 14**). Two commonly used ITS extraction tools are ITSx [38] and ITSxpress [39], and these are discussed below.
4. ITSx accepts query sequences in FASTA format [40], either with or without gaps, and has no restriction on the number of sequences that can be processed [38]. The software initially analyses the sequences in their default orientation and then repeats the search in the reverse complementary orientation to accommodate incorrectly oriented sequences. LotuS2 [27], PipeCraft2 [29], and FROGS [31] all utilize ITSx.
5. ITSxpress was developed to enhance the functionality of ITSx by extending its application to work with marker gene studies utilizing exact sequence variants (ESVs) [39, 41]. Sequence analysis methods, such as DADA2 [42] and Deblur [43], are frequently used to identify ESVs, which represent true biological sequences. These methods require trimming of FASTQ sequences to accurately infer error profiles. Unlike

ITSx, ITSxpress is specifically designed to merge and trim FASTQ sequences and is integrated into QIIME2 [28].

3.3.3 OTU Inference

ITS sequences can display significant length variation [24], making alignment-based clustering methods less suitable. Consequently, OTU inference approaches that account for this variability are more appropriate for ITS datasets, and while clustering tools are often optimized for 16S rRNA sequences, where alignment gaps are less frequent [24], adjusting gap and gap extension penalties to lower values can help improve clustering.

OTU inference methods commonly used for clustering both 16S rRNA (see Chap. 6) and ITS sequences include VSEARCH [44] and UPARSE [45], as well as modern clustering-free methods like DADA2 [42] and Deblur [43].

1. DADA2 [42] and Deblur [43] rely on error profile inference to resolve sequences into ESVs. However, these algorithms pose complications in the analysis of the ITS datasets without the use of ITSxpress since they require uniform sequence length. These approaches also perform poorly with ITS datasets because fungal genomes frequently display haplotype variation, where two or more distinct rRNA gene and ITS copies can co-exist within a single genome or haploid nuclei [25]. This phenomenon is particularly prevalent in dikaryotic genomes (e.g., Basidiomycota), diploid genomes, and polyploid genomes [26]. As a result, ESV methods tend to overestimate the richness of common fungal species while simultaneously underestimating the richness of rare species by filtering out rare variants.
2. While ESVs are not generally recommended for clustering ITS sequences, if they are used, then it is advisable to perform an additional clustering step to achieve species-level resolution [25].
3. CD-HIT is an effective OTU clustering method for ITS data due to its "word-based clustering approach," which divides sequences into fixed-length substrings (k-mers), rather than relying on full-length alignment [46], avoiding the challenges often associated with ITS sequence length variation. The LotuS2 pipeline [27] adopts CD-HIT [46] as its default clustering algorithm for ITS datasets, leveraging its efficiency and suitability for this type of data (*see* **Note 15**).

3.3.4 Taxonomic Assignment

Tools available include the following:

1. BLAST [47] represents the most commonly used method for analyzing amplicon reads [26]. In BLAST searches, a word size smaller than 10 is recommended to achieve long query-to-template alignments, enabling more accurate estimates of E-value and sequence similarity [26].

2. The Naïve Bayesian Classifier [48] and SINTAX [49] are up to 100 times faster than BLAST, which along with the alignment-based PROTAX-fungi [50] offer probabilistic estimates of taxonomic precision. PROTAX-fungi [50], integrated into the PlutoF platform [51] of the UNITE database [52], is the best-matching reference sequence for a query, and provides a comprehensive list of possible assignments with their associated probabilities, allowing the user to evaluate the reliability of the results.
3. The AMPtk amplicon analysis pipeline [30] adopts a hybrid approach for taxonomy assignment by generating a taxonomy table based on the consensus of multiple taxonomic assignment tools. While this method may result in more conservative assignments, it significantly enhances confidence in the final taxonomic classifications.
4. For taxonomic classification, a well-curated taxonomic database is essential to ensure accurate assignments.
5. The UNITE database (https://unite.ut.ee) serves as a curated public database of full-length ITS sequences, free from ambiguous entries [52]. It is currently the most comprehensive resource for taxonomic classification of ITS sequences, providing ready-to-use files compatible with various pipelines (e.g., LotuS2 [27] and QIIME2 [28]). The database currently contains 3,846,536 ITS sequence entries (as of April 2024, UNITE version 10.0).
6. The UNITE database is built on the "species hypothesis" (SH) concept, which defines taxa based on clustering at similarity thresholds of 97–99%, roughly corresponding to the species level [52]. Each SH is represented by a manually or automatically curated sequence and is linked to a unique, permanent digital object identifier (DOI). This DOI enables unambiguous identification of SHs, even in the absence of formal taxonomic names or when a fungal OTU has not been taxonomically classified.

3.3.5 Generating ITS-Derived Phylogenetic Trees

The significant length variation in the fungal ITS region makes accurate large-scale alignment of ITS sequences challenging [53], often resulting in poor alignments. These, in turn, lead to unreliable phylogenetic trees. Phylogenetic information plays a critical role in taxonomic placement of unknown sequences and calculation of phylogenetic diversity metrics (e.g., Faith's Phylogenetic Diversity index [54] and UniFrac [55]).

Users have two options regarding the phylogenetic analysis of ITS data:

1. Skip phylogenetic tree generation and rely on non-phylogenetic diversity metrics (e.g., Shannon Index or Bray-Curtis dissimilarity).
2. Construct phylogenetic trees using methods that account for the unique characteristics of ITS sequences.

While constructing phylogenetic trees from ITS sequence alignments is possible, it requires specialized software solutions. One such tool is *ghost-tree* [53], which integrates data from two marker genes into a single hybrid phylogenetic tree suitable for diversity analyses as follows:

1. A "foundation" tree is built using sequences from a conserved marker gene (e.g., 18S rRNA) that can be reliably aligned across distantly related taxa.
2. Sequences from a less conserved marker gene (e.g., the ITS region), which offers higher taxonomic resolution, are aligned within groups of closely related taxa to produce "extension" trees.
3. The extension trees are then grafted onto the foundation tree, creating the final "ghost tree" for downstream analyses.

Given the inherent challenges, the selection of analytical methods and parameter tuning must be approached on a case-by-case basis, accounting for both dataset quality and study objective(s). For instance, both ESV- and OTU-based clustering methods can produce comparable β diversity metrics, and the use of either may result in similar ecological interpretations [56]. However, using an ESV-based clustering method without additional clustering runs the risk of misrepresenting haplotypes as species, and overestimating fungal diversity [25]. Thus, careful methodological choices and thoughtful interpretation remain critical for robust and biologically meaningful ITS analyses.

4 Notes

1. Sabouraud dextrose (SD) and "Yeast extract-Malt extract" (YM) media are recommended selective growth media, as both are used routinely for the cultivation of yeast and other fungi from a variety of sources, including human fecal samples. In addition, each growth medium should be supplemented with antibiotics to suppress/inhibit bacterial overgrowth. The addition of two antibiotics, rather than just one, works more effectively to inhibit bacterial growth. Rose Bengal (RB) agar is another selective medium that can be used for fungal cultivation. However, lipophilic fungi (e.g., *Malassezia* spp.) do not

grow well, if at all, on these and require specialized media (e.g., modified Dixon's agar).

2. When weighing out feces, the quantity used is dependent upon the original amount provided by the study participant (e.g., infant vs adult).
3. Inoculation of all solid and liquid media should be carried out aseptically either in a safety cabinet or next to a lit Bunsen burner flame.
4. Cells of *Saccharomyces cerevisiae*, a food-associated yeast often present within the human gut [1, 2], measure between 5 and 10 μm in length [20]. In contrast, cells of *Bacteroides thetaiotamicron*, one of the most common bacterial species in the human gut microbiota, typically measure 1 μm.
5. Given that some bacterial species can acquire antibiotic resistance and may grow on antibiotic-supplemented agar, it is recommended that representative suspect colonies are examined using a standard light microscope. Yeast and filamentous fungi can often be distinguished from bacteria based on their larger cell size.
6. For identification (and cultivation) of candidate gut fungal commensals (e.g., *Candida albicans*), incubation is at 37 °C (i.e., host core body temperature), rather than at 30 °C.
7. An empty bead-beating tube should be included as an extraction control and treated the same as for the tubes containing fecal samples. This control represents the "kitome" and is used to identify DNA contaminants that may be introduced during the DNA extraction process (e.g., from reagents).
8. When dispensing fecal homogenate, the use of a wide-bore pipette tip may help, especially if the homogenate still contains particulate matter post vortexing.
9. Typical DNA yields from adult human fecal samples using the above extraction protocol often exceed 100 ng/μL, making the Qubit dsDNA Broad-Range (BR) assay kit ideally suited for accurate quantification.
10. It is good practice to use both positive and negative controls in all PCR runs and DNA extractions. The inclusion of a negative control can help identify any contaminants that may be inadvertently introduced during processing, from reagents or from the DNA extraction kit itself.
11. For human fecal DNA samples, use 100 ng of DNA as template, diluted to a final working concentration of 10 ng/μL. This amount is usually sufficient for successful fungal DNA amplification from human fecal DNA samples.

12. When doing database searches to either determine or confirm species identity, it is essential to ensure any comparisons are made against authenticated reference strains (i.e., species type strains). For instance, when using BLASTN for such purposes to search rRNA/ITS databases (https://blast.ncbi.nlm.nih.gov/Blast.cgi), the "Sequences from type material" option should always be selected. For ITS-based identification, strains of the same species typically display sequence similarity values of 97% or more [52]. However, there are exceptions (e.g., *Clavispora lusitaniae* [57]), meaning that ITS sequence similarity values should be used as a preliminary diagnostic guide to determining species identity [58].
13. Midori Green Direct DNA replaces toxic ethidium bromide and allows DNA to be visualized under blue light using a blue/green LED transilluminator.
14. Not all pipelines include an option for ITS extraction (e.g., DADA2), and this can significantly impact the overall accuracy of ITS dataset processing. The importance of this step was first highlighted using the LotuS2 pipeline [27]. This pipeline, like QIIME2 [28], implements ITS extraction and demonstrated superior reproducibility of fungal composition compared to DADA2.
15. When OTU inference methods were benchmarked in the LotuS2 pipeline using soil sample-derived ITS datasets, CD-HIT [46] and UPARSE [45] showed the best reproducibility of fungal composition among technical replicates.

Acknowledgements

This work was funded by the Biotechnology and Biological Sciences Research Council (BBSRC) Institute Strategic Programme Food Microbiome and Health BB/X011054/1 and its constituent projects, by the Institute Strategic Programme Grant Gut Microbes and Health BB/R012490/1 and its constituent project, and by the BBSRC Core Capability Grant BB/CCG1860/1.

References

1. Hallen-Adams HE, Suhr MJ (2017) Fungi in the healthy human gastrointestinal tract. Virulence 8(3):352–358. https://doi.org/10.1080/21505594.2016.1247140
2. Nash AK, Auchtung TA, Wong MC, Smith DP, Gesell JR, Ross MC, Stewart CJ, Metcalf GA, Muzny DM, Gibbs RA, Ajami NJ, Petrosino JF (2017) The gut mycobiome of the human microbiome project healthy cohort. Microbiome 5:153. https://doi.org/10.1186/s40168-017-0373-4
3. Belvoncikova P, Splichalova P, Videnska P, Gardlik R (2022) The human mycobiome: colonization, composition and the role in health and disease. J Fungi 8(10):1046. https://doi.org/10.3390/jof8101046

4. Qin JJ, Li RQ, Raes J, Arumugam M, Burgdorf KS, Manichanh C, Nielsen T, Pons N, Levenez F, Yamada T, Mende DR, Li JH, Xu JM, Li SC, Li DF, Cao JJ, Wang B, Liang HQ, Zheng HS, Xie YL, Tap J, Lepage P, Bertalan M, Batto JM, Hansen T, Le Paslier D, Linneberg A, Nielsen HB, Pelletier E, Renault P, Sicheritz-Ponten T, Turner K, Zhu HM, Yu C, Li ST, Jian M, Zhou Y, Li YR, Zhang XQ, Li SG, Qin N, Yang HM, Wang J, Brunak S, Dore J, Guarner F, Kristiansen K, Pedersen O, Parkhill J, Weissenbach J, Bork P, Ehrlich SD, Wang J, Meta HITC (2010) A human gut microbial gene catalogue established by metagenomic sequencing. Nature 464(7285): 59–70. https://doi.org/10.1038/nature08821
5. Arumugam M, Raes J, Pelletier E, Le Paslier D, Yamada T, Mende DR, Fernandes GR, Tap J, Bruls T, Batto JM, Bertalan M, Borruel N, Casellas F, Fernandez L, Gautier L, Hansen T, Hattori M, Hayashi T, Kleerebezem M, Kurokawa K, Leclerc M, Levenez F, Manichanh C, Nielsen HB, Nielsen T, Pons N, Poulain J, Qin JJ, Sicheritz-Ponten T, Tims S, Torrents D, Ugarte E, Zoetendal EG, Wang J, Guarner F, Pedersen O, de Vos WM, Brunak S, Doré J, Weissenbach J, Ehrlich SD, Bork P, Kristiansen K, Meta HITC (2011) Enterotypes of the human gut microbiome (vol 473, pg 174, 2011). Nature 474 (7353):174-180. https://doi.org/10.1038/nature10187
6. Underhill DM, Braun J (2022) Fungal microbiome in inflammatory bowel disease: a critical assessment. J Clin Invest 132(5):e155786. https://doi.org/10.1172/jci155786
7. Li XV, Leonardi I, Iliev ID (2019) Gut mycobiota in immunity and inflammatory disease. Immunity 50(6):1365–1379. https://doi.org/10.1016/j.immuni.2019.05.023
8. Zhang F, Aschenbrenner D, Yoo JY, Zuo T (2022) The gut mycobiome in health, disease, and clinical applications in association with the gut bacterial microbiome assembly. Lancet Microbe 3(12):E969–E983. https://doi.org/10.1016/s2666-5247(22)00203-8
9. Sokol H, Leducq V, Aschard H, Pham HP, Jegou S, Landman C, Cohen D, Liguori G, Bourrier A, Nion-Larmurier I, Cosnes J, Seksik P, Langella P, Skurnik D, Richard ML, Beaugerie L (2017) Fungal microbiota dysbiosis in IBD. Gut 66(6):1039–1048. https://doi.org/10.1136/gutjnl-2015-310746
10. Gosiewski T, Salamon D, Szopa M, Sroka A, Malecki MT, Bulanda M (2014) Quantitative evaluation of fungi of the genus *Candida* in the feces of adult patients with type 1 and 2 diabetes – a pilot study. Gut Pathogens 6(1):43. https://doi.org/10.1186/s13099-014-0043-z
11. Luan CG, Xie LL, Yang X, Miao HF, Lv N, Zhang RF, Xiao X, Hu YF, Liu YL, Wu N, Zhu YM, Zhu BL (2015) Dysbiosis of fungal microbiota in the intestinal mucosa of patients with colorectal adenomas. Sci Rep 5:7980. https://doi.org/10.1038/srep07980
12. Rodriguez MM, Perez D, Chaves FJ, Esteve E, Marin-Garcia P, Xifra G, Vendrell J, Jove M, Pamplona R, Ricart W, Portero-Otin M, Chacon MR, Real JMF (2015) Obesity changes the human gut mycobiome. Sci Rep 5:14600. https://doi.org/10.1038/srep14600
13. Botschuijver S, Roeselers G, Levin E, Jonkers DM, Welting O, Heinsbroek SEM, de Weerd HH, Boekhout T, Fornai M, Masclee AA, Schuren FHJ, de Jonge WJ, Seppen J, Van den Wijngaard RM (2017) Intestinal fungal dysbiosis is associated with visceral hypersensitivity in patients with irritable bowel syndrome and rats. Gastroenterology 153(4): 1026–1039. https://doi.org/10.1053/j.gastro.2017.06.004
14. Coker OO, Nakatsu G, Dai RZ, Wu WKK, Wong SH, Ng SC, Chan FKL, Sung JJY, Yu J (2019) Enteric fungal microbiota dysbiosis and ecological alterations in colorectal cancer. Gut 68(4):654–662. https://doi.org/10.1136/gutjnl-2018-317178
15. Jayasudha R, Das T, Chakravarthy SK, Prashanthi GS, Bhargava A, Tyagi M, Rani PK, Pappuru RR, Shivaji S (2020) Gut mycobiomes are altered in people with type 2 diabetes mellitus and diabetic retinopathy. PLoS One 15(12). https://doi.org/10.1371/journal.pone.0243077
16. James SA, O'Kelly MJT, Carter DM, Davey RP, van Oudenaarden A, Roberts IN (2009) Repetitive sequence variation and dynamics in the ribosomal DNA array of *Saccharomyces cerevisiae* as revealed by whole-genome resequencing. Genome Res 19(4):626–635. https://doi.org/10.1101/gr.084517.108
17. Auchtung TA, Fofanova TY, Stewart CJ, Nash AK, Wong MC, Gesell JR, Auchtung JM, Ajami NJ, Petrosino JF (2018) Investigating colonization of the healthy adult gastrointestinal tract by fungi. Msphere 3(2):e00092-18. https://doi.org/10.1128/mSphere.00092-18
18. James SA, Phillips S, Telatin A, Baker D, Ansorge R, Clarke P, Hall LJ, Carding SR (2020) Preterm infants harbour a rapidly changing mycobiota that includes *Candida*

pathobionts. J Fungi 6(4):273. https://doi.org/10.3390/jof6040273

19. Huseyin CE, Rubio RC, O'Sullivan O, Cotter PD, Scanlan PD (2017) The fungal frontier: a comparative analysis of methods used in the study of the human gut mycobiome. Front Microbiol 8:1432. https://doi.org/10.3389/fmicb.2017.01432
20. Vaughan-Martini A, Martini A (2011) *Saccharomyces* Meyen & Reess (1870). In: Kurtzman CP, Fell JW, Boekhout T (eds) The yeasts; a taxonomic study, vol 2, 5th edn. Elsevier, Burlington, pp 733–746
21. White TJ, Bruns TD, Lee SL, Taylor JW (1990) Amplification and direct sequencing of fungal ribosomal RNA genes for phylogenetics. In: Innis MA, Gelfand DH, Sninsky JJ (eds) PCR protocols: a guide to methods and applications. Academic Press, San Diego, pp 315–322
22. Gardes M, Bruns TD (1993) ITS primers with enhanced specificity for basidiomycetes—application to the identification of mycorrhizae and rusts. Mol Ecol 2(2):113–118. https://doi.org/10.1111/j.1365-294x.1993.tb00005.x
23. Motooka D, Fujimoto K, Tanaka R, Yaguchi T, Gotoh K, Maeda Y, Furuta Y, Kurakawa T, Goto N, Yasunaga T, Narazaki M, Kumanogoh A, Horii T, Iida T, Takeda K, Nakamura S (2017) Fungal ITS1 deep-sequencing strategies to reconstruct the composition of a 26-species community and evaluation of the gut mycobiota of healthy Japanese individuals. Front Microbiol 8:238. https://doi.org/10.3389/fmicb.2017.00238
24. Nilsson R, Anslan S, Bahram M, Wurzbacher C, Baldrian P, Tedersoo L (2019) Mycobiome diversity: high-throughput sequencing and identification of fungi. Nat Rev Microbiol 17(2):95–109. https://doi.org/10.1038/s41579-018-0116-y
25. Estensmo E, Maurice S, Morgado L, Martin-Sanchez P, Skrede I, Kauserud H (2021) The influence of intraspecific sequence variation during DNA metabarcoding: a case study of eleven fungal species. Mol Ecol Resour 21(4):1141–1148. https://doi.org/10.1111/1755-0998.13329
26. Tedersoo L, Bahram M, Zinger L, Nilsson R, Kennedy P, Yang T, Anslan S, Mikryukov V (2022) Best practices in metabarcoding of fungi: from experimental design to results. Mol Ecol 31(10):2769–2795. https://doi.org/10.1111/mec.16460
27. Özkurt E, Fritscher J, Soranzo N, Ng D, Davey R, Bahram M, Hildebrand F (2022) LotuS2: an ultrafast and highly accurate tool for amplicon sequencing analysis. Microbiome 10(1):176. https://doi.org/10.1186/s40168-022-01365-1
28. Bolyen E, Rideout JR, Dillon MR, Bokulich NA, Abnet CC, Al-Ghalith GA, Alexander H, Alm EJ, Arumugam M, Asnicar F, Bai Y, Bisanz JE, Bittinger K, Brejnrod A, Brislawn CJ, Brown CT, Callahan BJ, Caraballo-Rodriguez AM, Chase J, Cope EK, Da Silva R, Diener C, Dorrestein PC, Douglas GM, Durall DM, Duvallet C, Edwardson CF, Ernst M, Estaki M, Fouquier J, Gauglitz JM, Gibbons SM, Gibson DL, Gonzalez A, Gorlick K, Guo J, Hillmann B, Holmes S, Holste H, Huttenhower C, Huttley GA, Janssen S, Jarmusch AK, Jiang LJ, Kaehler BD, Bin Kang K, Keefe CR, Keim P, Kelley ST, Knights D, Koester I, Kosciolek T, Kreps J, Langille MGI, Lee J, Ley R, Liu YX, Loftfield E, Lozupone C, Maher M, Marotz C, Martin BD, McDonald D, McIver LJ, Melnik AV, Metcalf JL, Morgan SC, Morton JT, Naimey AT, Navas-Molina JA, Nothias LF, Orchanian SB, Pearson T, Peoples SL, Petras D, Preuss ML, Pruesse E, Rasmussen LB, Rivers A, Robeson MS, Rosenthal P, Segata N, Shaffer M, Shiffer A, Sinha R, Song SJ, Spear JR, Swafford AD, Thompson LR, Torres PJ, Trinh P, Tripathi A, Turnbaugh PJ, Ul-Hasan S, van der Hooft JJJ, Vargas F, Vazquez-Baeza Y, Vogtmann E, von Hippel M, Walters W, Walters W, Wan Y, Wang M, Warren J, Weber KC, Williamson CHD, Willis AD, Xu ZZ, Zaneveld JR, Zhang YL, Zhu QY, Knight R, Caporaso JG (2019) Reproducible, interactive, scalable and extensible microbiome data science using QIIME 2 (vol 37, pg 852, 2019). Nat Biotechnol 37(9):1091–1091. https://doi.org/10.1038/s41587-019-0252-6
29. Anslan S, Bahram M, Hiiesalu I, Tedersoo L (2017) PipeCraft: flexible open-source toolkit for bioinformatics analysis of custom high-throughput amplicon sequencing data. Mol Ecol Resour 17(6):e234–e240. https://doi.org/10.1111/1755-0998.12692
30. Palmer J, Jusino M, Banik M, Lindner D (2018) Non-biological synthetic spike-in controls and the AMPtk software pipeline improve mycobiome data. PeerJ 6:e4925. https://doi.org/10.7717/peerj.4925
31. Bernard M, Rué O, Mariadassou M, Pascal G (2021) FROGS: a powerful tool to analyse the diversity of fungi with special management of internal transcribed spacers. Brief Bioinform 22(6):1-6. https://doi.org/10.1093/bib/bbab318
32. Loos D, Zhang L, Beemelmanns C, Kurzai O, Panagiotou G (2021) DAnIEL: a user-friendly

web server for fungal ITS amplicon sequencing data. Front Microbiol 12:720513. https://doi.org/10.3389/fmicb.2021.720513

33. Gweon H, Oliver A, Taylor J, Booth T, Gibbs M, Read D, Griffiths R, Schonrogge K (2015) PIPITS: an automated pipeline for analyses of fungal internal transcribed spacer sequences from the Illumina sequencing platform. Methods Ecol Evol 6(8):973–980. https://doi.org/10.1111/2041-210X.12399
34. Hakimzadeh A, Asbun A, Albanese D, Bernard M, Buchner D, Callahan B, Caporaso J, Curd E, Djemiel C, Durling M, Elbrecht V, Gold Z, Gweon H, Hajibabaei M, Hildebrand F, Mikryukov V, Normandeau E, Özkurt E, Palmer J, Pascal G, Porter T, Straub D, Vasar M, Vetrovsky T, Zafeiropoulos H, Anslan S (2024) A pile of pipelines: an overview of the bioinformatics software for metabarcoding data analyses. Mol Ecol Resour 24(5):e13847. https://doi.org/10.1111/1755-0998.13847
35. Ansorge R, Birolo G, James SA, Telatin A (2021) Dadaist2: a toolkit to automate and simplify statistical analysis and plotting of metabarcoding experiments. Int J Mol Sci 22(10):5309. https://doi.org/10.3390/ijms22105309
36. Asbun A, Besseling M, Balzano S, van Bleijswijk J, Witte H, Villanueva L, Engelmann J (2020) Cascabel: a scalable and versatile amplicon sequence data analysis pipeline delivering reproducible and documented results. Front Genet 11:489357. https://doi.org/10.3389/fgene.2020.489357
37. Halwachs B, Madhusudhan N, Krause R, Nilsson R, Moissl-Eichinger C, Högenauer C, Thallinger G, Gorkiewicz G (2017) Critical issues in mycobiota analysis. Front Microbiol 8:180. https://doi.org/10.3389/fmicb.2017.00180
38. Bengtsson-Palme J, Ryberg M, Hartmann M, Branco S, Wang Z, Godhe A, De Wit P, Sánchez-García M, Ebersberger I, de Sousa F, Amend A, Jumpponen A, Unterseher M, Kristiansson E, Abarenkov K, Bertrand Y, Sanli K, Eriksson K, Vik U, Veldre V, Nilsson R (2013) Improved software detection and extraction of ITS1 and ITS2 from ribosomal ITS sequences of fungi and other eukaryotes for analysis of environmental sequencing data. Methods Ecol Evol 4(10):914–919. https://doi.org/10.1111/2041-210X.12073
39. Rivers A, Weber K, Gardner T, Liu S, Armstrong S (2018) ITSxpress: software to rapidly trim internally transcribed spacer sequences with quality scores for marker gene analysis [version 1; peer review: 2 approved]. F1000Research 7:1418. https://doi.org/10.12688/f1000research.15704.1
40. Pearson W, Lipman D (1988) Improved tools for biological sequence comparison. Proc Natl Acad Sci USA 85(8):2444–2448. https://doi.org/10.1073/pnas.85.8.2444
41. Einarsson S, Rivers A (2024) ITSxpress version 2: software to rapidly trim internal transcribed spacer sequences with quality scores for amplicon sequencing. Microbiol Spectr 12(12):e0060124. https://doi.org/10.1128/spectrum.00601-24
42. Callahan BJ, McMurdie PJ, Rosen MJ, Han AW, Johnson AJA, Holmes SP (2016) DADA2: high-resolution sample inference from Illumina amplicon data. Nat Methods 13(7):581-583. https://doi.org/10.1038/nmeth.3869
43. Amir A, McDonald D, Navas-Molina J, Kopylova E, Morton J, Xu Z, Kightley E, Thompson L, Hyde E, Gonzalez A, Knight R (2017) Deblur rapidly resolves single-nucleotide community sequence patterns. mSystems 2(2):e00191-16. https://doi.org/10.1128/mSystems.00191-16
44. Rognes T, Flouri T, Nichols B, Quince C, Mahé F (2016) VSEARCH: a versatile open source tool for metagenomics. PeerJ 4:e2584. https://doi.org/10.7717/peerj.2584
45. Edgar R (2013) UPARSE: highly accurate OTU sequences from microbial amplicon reads. Nat Methods 10(10):996-998. https://doi.org/10.1038/NMETH.2604
46. Fu L, Niu B, Zhu Z, Wu S, Li W (2012) CD-HIT: accelerated for clustering the next-generation sequencing data. Bioinformatics 28(23):3150–3152. https://doi.org/10.1093/bioinformatics/bts565
47. Camacho C, Coulouris G, Avagyan V, Ma N, Papadopoulos J, Bealer K, Madden T (2009) BLAST plus: architecture and applications. BMC Bioinformatics 10:421. https://doi.org/10.1186/1471-2105-10-421
48. Porras-Alfaro A, Liu K, Kuske C, Xie G (2014) From genus to phylum: large-subunit and internal transcribed spacer rRNA operon regions show similar classification accuracies influenced by database composition. Appl Environ Microbiol 80(3):829–840. https://doi.org/10.1128/AEM.02894-13
49. Edgar RC (2016) SINTAX: a simple non-Bayesian taxonomy classifier for 16S and ITS sequences. bioRxiv:074161. https://doi.org/10.1101/074161
50. Abarenkov K, Somervuo P, Nilsson R, Kirk P, Huotari T, Abrego N, Ovaskainen O (2018) Protax-fungi: a web-based tool for probabilistic

taxonomic placement of fungal internal transcribed spacer sequences. New Phytol 220(2): 517–525. https://doi.org/10.1111/nph.15301

51. Abarenkov K, Tedersoo L, Nilsson R, Vellak K, Saar I, Veldre V, Parmasto E, Prous M, Aan A, Ots M, Kurina O, Ostonen I, Jogeva J, Halapuu S, Poldmaa K, Toots M, Truu J, Larsson K, Koljalg U (2010) PlutoF—a web based workbench for ecological and taxonomic research, with an online implementation for fungal ITS sequences. Evol Bioinforma 6: 189–196. https://doi.org/10.4137/EBO.S6271
52. Kõljalg U, Nilsson R, Abarenkov K, Tedersoo L, Taylor A, Bahram M, Bates S, Bruns T, Bengtsson-Palme J, Callaghan T, Douglas B, Drenkhan T, Eberhardt U, Dueñas M, Grebenc T, Griffith G, Hartmann M, Kirk P, Kohout P, Larsson E, Lindahl B, Luecking R, Martin M, Matheny P, Nguyen N, Niskanen T, Oja J, Peay K, Peintner U, Peterson M, Poldmaa K, Saag L, Saar I, Schüessler A, Scott J, Senes C, Smith M, Suija A, Taylor D, Telleria M, Weiss M, Larsson K (2013) Towards a unified paradigm for sequence-based identification of fungi. Mol Ecol 22(21):5271–5277. https://doi.org/10.1111/mec.12481
53. Fouquier J, Rideout J, Bolyen E, Chase J, Shiffer A, McDonald D, Knight R, Caporaso J, Kelley S (2016) Ghost-tree: creating hybrid-gene phylogenetic trees for diversity analyses. Microbiome 4:11. https://doi.org/10.1186/s40168-016-0153-6
54. Faith D (1992) Conservation evaluation and phylogenetic diversity. Biol Conserv 61(1): 1–10. https://doi.org/10.1016/0006-3207(92)91201-3
55. Lozupone C, Knight R (2005) UniFrac: a new phylogenetic method for comparing microbial communities. Appl Environ Microbiol 71(12): 8228–8235. https://doi.org/10.1128/AEM.71.12.8228-8235.2005
56. Glassman S, Martiny J (2018) Broadscale ecological patterns are robust to use of exact sequence variants versus operational taxonomic units. mSphere 3(4):e00148-18. https://doi.org/10.1128/mSphere.00148-18
57. Lachance M, Daniel H, Meyer W, Prasad G, Gautam S, Boundy-Mills K (2003) The D1/D2 domain of the large-subunit rDNA of the yeast species *Clavispora lusitaniae* is unusually polymorphic. FEMS Yeast Res 4(3): 253–258. https://doi.org/10.1016/S1567-1356(03)00113-2
58. Lücking R, Aime M, Robbertse B, Miller A, Ariyawansa H, Aoki T, Cardinali G, Crous P, Druzhinina I, Geiser D, Hawksworth D, Hyde K, Irinyi L, Jeewon R, Johnston P, Kirk P, Malosso E, May T, Meyer W, Öpik M, Robert V, Stadler M, Thines M, Vu D, Yurkov A, Zhang N, Schoch C (2020) Unambiguous identification of fungi: where do we stand and how accurate and precise is fungal DNA barcoding? IMA Fungus 11(1):14. https://doi.org/10.1186/s43008-020-00033-z

INDEX

B

Simon R. Carding (ed.), *Best Practice in Microbiome Research*, Springer Protocols Handbooks,
https://doi.org/10.1007/978-1-0716-5009-7, © The Editor(s) (if applicable) and The Author(s) 2026

K

M

N

O

R

S

MIX
Papier aus verantwortungsvollen Quellen
Paper from responsible sources
FSC® C105338

If you have any concerns about our products,
you can contact us on
ProductSafety@springernature.com

In case Publisher is established outside the EU,
the EU authorized representative is:
Springer Nature Customer Service Center GmbH
Europaplatz 3, 69115 Heidelberg, Germany

Printed by Libri Plureos GmbH
in Hamburg, Germany